FESTOOL® ESSENTIALS

TS 55 EQ & TS 75 EQ Portable Plunge Saws

with FS/2 Guide Rail, MFT Multi-function Table, & CT Dust Extraction Systems

Schiffer® Publishing Ltd

4880 Lower Valley Road Atglen, Pennsylvania 19310

Schiffer Books are available at special discounts for bulk purchases for sales promotions or premiums. Special editions, including personalized covers, corporate imprints, and excerpts can be created in large quantities for special needs. For more information contact the publisher:

Published by Schiffer Publishing Ltd.
4880 Lower Valley Road
Atglen, PA 19310
Phone: (610) 593-1777; Fax: (610) 593-2002
E-mail: Info@schifferbooks.com

For the largest selection of fine reference books on this and related subjects, please visit our web site at **www.schifferbooks.com**
We are always looking for people to write books on new and related subjects. If you have an idea for a book please contact us at the above address.

This book may be purchased from the publisher.
Include $5.00 for shipping.
Please try your bookstore first.
You may write for a free catalog.

In Europe, Schiffer books are distributed by
Bushwood Books
6 Marksbury Ave.
Kew Gardens
Surrey TW9 4JF England
Phone: 44 (0) 20 8392-8585; Fax: 44 (0) 20 8392-9876
E-mail: info@bushwoodbooks.co.uk
Website: www.bushwoodbooks.co.uk
Free postage in the U.K., Europe; air mail at cost.

Library of Congress Control Number: 2008930843

Covers and book designed by: Bruce Waters
Type set in Huminist 521

ISBN: 978-0-7643-3103-9
Printed in China

CONTENTS

INTRODUCTION

The company Albert Fezer and Gottlieb Stoll founded in 1925, in southern Germany, has been known by many names, Fezer & Stoll, Festo, Festo Tooltechnic, and, beginning in 2000, Festool. Whatever the name, the mission has always been the same: to provide expert craftspeople with the finest tools available to make their work more efficient, easier, and better. From the beginning they have committed themselves to understanding the needs of the craftsperson in an everchanging workplace. They have addressed those needs with innovation, quality, and performance that is unsurpassed in the industry. The result is captured in their three word credo: "Faster, Easier, Smarter."

Today the name Festool stands for an integrated system of tools engineered for ease of use and reliability, to increase the accuracy, creativity, and productivity of the workplace. Each tool in designed to be used with the powerful dust extraction system, which is covered here in great detail.

Used with Festool plunge saws and routers, the Festool guide and work table systems allow for unsurpassed accuracy, convenience, and flexibility.

1. THE SAWS

The Festool system brings an unexcelled level of precision to the portable circular saw. The TS 55 EQ and TS 75 EQ plunge saws, when used with the guide rail system, provide the precision of stationary tools while keeping the versatility and portability of traditional circular saws. With the guide rail and its splinter guards the cuts are virtually splinter-free.

They are also straight. The guide rail system eliminates the tendency of straying off the line, whether with a perpendicular cut or a bevel. The engineering of these saws keeps the cut line the same for either one. Also, as a plunge saw, it is possible to start or end a cut accurately at any point in the material.

The saws are also safer than the conventional circular saw. The saw blade and riving knife retract into the housing as the saw is lifted from the material.

Main Features of the Festool Portable Plunge Saws

• **Zero play guidance.** Tool-less adjustment of the alignment cams on the guide rail for accurate, perfectly straight sawing.

• **Scribe line = cut line.** Exact cuts, even at an angle. The TS 55 EQ and TS 75 EQ make the actual cut right on the edge of the scribe line, even on bevel cuts.

• **Splinterguard on both sides of the cut.** Splinter-free cuts on both sides of the cut line. The attachable splinterguard and the guide rail make this possible.

• **Spring-loaded riving knife or splitter.** Helps to prevent the binding of the saw blade by acting as a wedge and keeping the kerf of the cut open. It retracts into the housing during plunge cuts

• **Triple motor bearing.** The triple motor bearing improves the smoothness of cutting action and extends the service life.

• **MMC electronics.** Equipped with variable speed for improved cut results by matching speed to blade type and material. Maximum adaptation to the working material, controlled speed and temperature monitoring.

• **FastFix® saw blade changing system.** The saw blade can be changed in just a few simple steps. This means you can change the blade more easily to suit the material you are working on, which improves cut performance for less material waste thus saving you money.

• **Rotating dust port.** More freedom of movement; fewer dangerous obstacles. The suction hose never gets tangled or in the way and can always remain attached for maximum dust collection.

• **Slip clutch (TS 75 EQ only).** The sliding clutch minimizes saw kickback should the saw blade bind, increasing safety.

• **Kickback stop.** Eliminates kickbacks. A kickback stop..engages the guide rail to reduce kickback and lift off during plunge cuts.

• **Easy depth adjustment in mm increments.** Indexed scale with stepped detents for precise depth setting for controlled and repeated cuts.

Portable Plunge Cut Saws TS 55 EQ & TS 75 EQ

Specifications

Model Name Item	TS 55 EQ 561 174	TS 75 EQ 561 188
Standard package includes		
High-quality carbide blade	48-tooth (495 377)	36-tooth (493 198)
Guide Rail included	55" (1400mm)	75" (1900mm)
Slipclutch		•
Splinterguard	•	•
Limit stop	•	•
Plug-it power cord	•	•
Systainer	SYS 4	SYS 5
Specifications		
Power consumption	1200 watts 10 amps 120 v AC	1600 watts 13 amps 120 v AC
Saw blade speed	2,000 - 5,200 rpm	1,350 - 3,550 rpm
Saw blade diameter	6-1/4" (160mm)	8-1/4" (210mm)
Bevel cuts	0° - 45°	0° - 45°
Cutting depth on guide rail	1-15/16" (50mm) straight down and 1-7/16" (37mm) at 45°	2-3/4" (70mm) straight down and 2-1/8" (55mm) at 45°
Cutting depth without guide rail	2-1/8" (55mm) straight down	2-15/16" (75mm) straight down
Arbor Size	.787" (20mm)	1.101" (30mm)
Weight	9.92 lbs. (4.5 kg)	13.6 lbs. (6.2 kg)

Accessories

Model Name	TS 55 EQ	TS 75 EQ
Parallel guide *for guided cuts, in parallel with indexing from an edge.*	491 469	492 243
Cover plate *encloses blade housing for improved dust extraction.*	491 750	
Splinterguard *replacement (pack of 5).*	491 473	491 473
Limit stop *helps reduce the risk of kickbacks. The saw comes with one, and a second limit stop can be used to set the limit on cuts (not just plunge cuts).*	491 582	491 582•
Plug-it power cord, *replacement.*	490 656	490 656

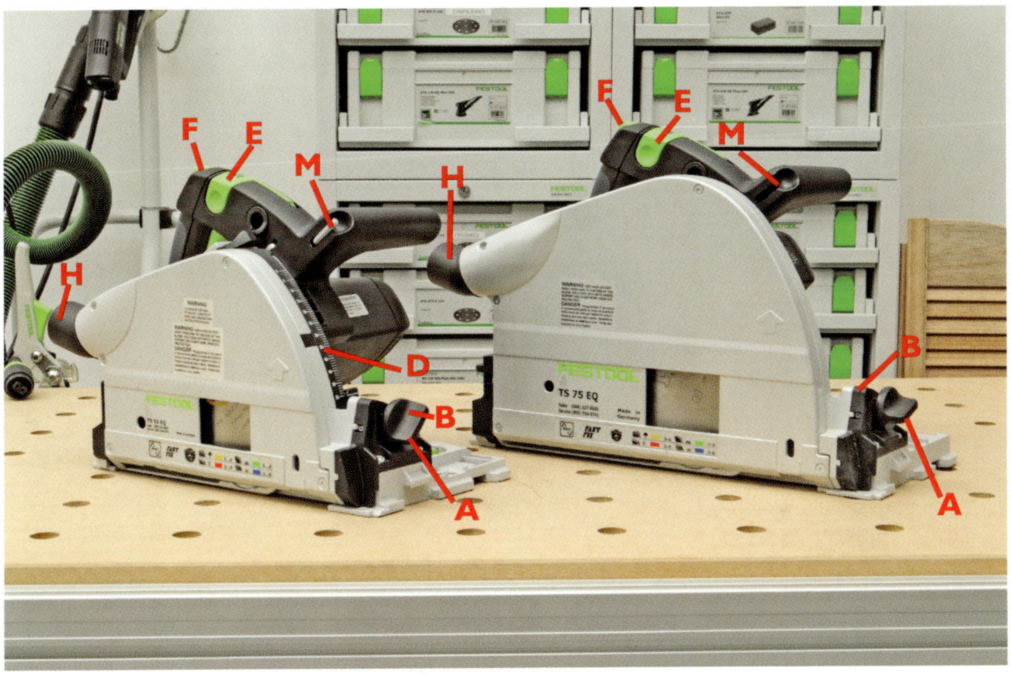

The TS 55 EQ and TS 75 EQ saws.

The motor side. The two saws both are products of advanced engineering. The main difference is, obviously, their size.

A: Thumb screw for angle adjustment

B: Angle scale

C: Depth stop for cutting depth adjustment

D. Cutting depth scale

E: FastFix lever for blade change

F: Switching on and unlocking block

G: On/off trigger switch

H: Dust exhaust port

I: Variable speed control dial

J: Power socket

K: Foot plate

L: Guide jaws

M: Hex wrench storage

1. THE SAWS

• The TS 55 EQ has a 1200 watt/10 amp motor; the TS 75 EQ has a 1600 watt/13 amp motor.

• The TS 55 EQ uses a 6-1/4"/160mm saw blade with a 20mm bore; the TS 75 EQ uses a 8-1/4"/210mm saw blade with a 30mm bore.

• Because of the difference in blade size, the TS 55 EQ has an rpm range of 2,000-5,200; the TS 75 range is 1,350-3,550 rpm.

• While the saws can be used off the Guide Rail system, they are designed to be integral to it. Cutting depth on the guide rail for the TS 55 EQ is 1-15/16" (50mm) at 90 degrees, and 1-7/16" (37mm)at 45 degrees; for the TS 75 EQ it is 2-3/4" (70mm) at 90 degrees and 2-1/8" (55mm) at 45 degrees.

The only engineering difference between the TS 55 EQ and the TS 75 EQ is that the TS 75 EQ has slip clutch. The motors on both saws will shut off when they are overheated. But the purpose of the slip clutch is to disengage the blade from the arbor when it encounters resistance in the kerf, preventing serious damage to the gearing system. This could be embedded material like a nail or such, that are common in reclaimed materials, or it could be an overly aggressive feed rate.

The saws are plunging saws. The blade retracts into the housing and is not exposed during cutting. This is a very important safety feature. Behind the blade is a spring-loaded riving knife, a feature that prevents binding of the blade.

The plunging action is initiated by depressing this button on the top of the handle. It is not necessary to hold this button after the blade is engaged.

The depth gauge is indexed in millimeters. It is set with the spring loaded indicator. Simply depress and move to the desired depth.

To set the blade easily, place the guide rail and the saw on the edge of the material to be cut. Plunge the saw until it touches the work table.

1. THE SAWS

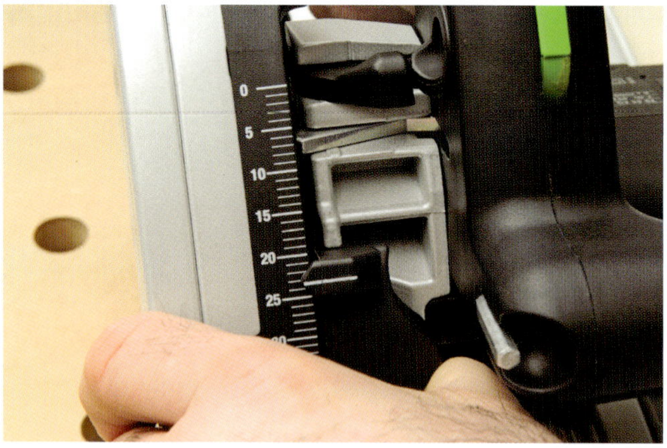

Holding the blade in place, slide the depth indicator up until it hits the stop on the saw casing.

Add 1 or 2 millimeters to the depth to insure that the blade goes through the material.

To set a bevel loosen the two locking knobs and adjust to the desired angle. Lock in place.

The two screws can be used to zero out the angle of the blade, should the need arise. One is at the lower front corner of the inner saw housing…

…the other is at the lower back corner.

Loosen the indicator lock with a 15mm torx, adjust the screws to zero the saw out, and retighten the lock screws.

The power cord is attached behind the motor housing and the vacuum hose connects to a swivel socket at the back of the blade housing. The 27mm hose fits inside the socket, the 36mm hose fits around it.

Blade Change

For blade changes, pull out the FastFix® lever to the upright position before plunging to lock out the depth position for easy access to the arbor nut. With the Fast Fix engaged the power switch is locked in the off position so I can safely change the blade.

Plunge the saw.

Blade Change

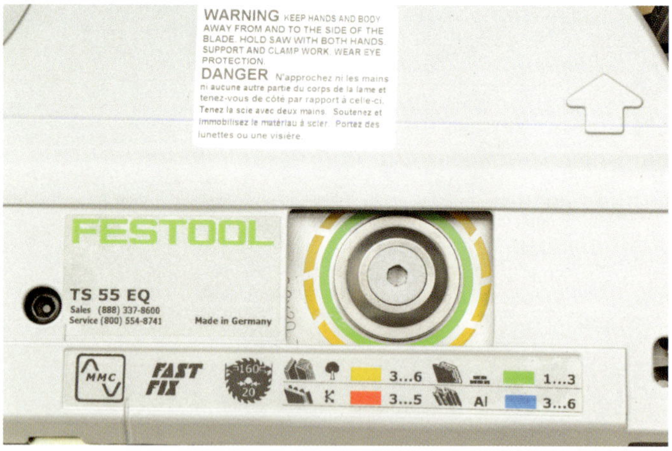

The arbor nut is now exposed through the housing window.

Remove the hex key from its storage place in the housing.

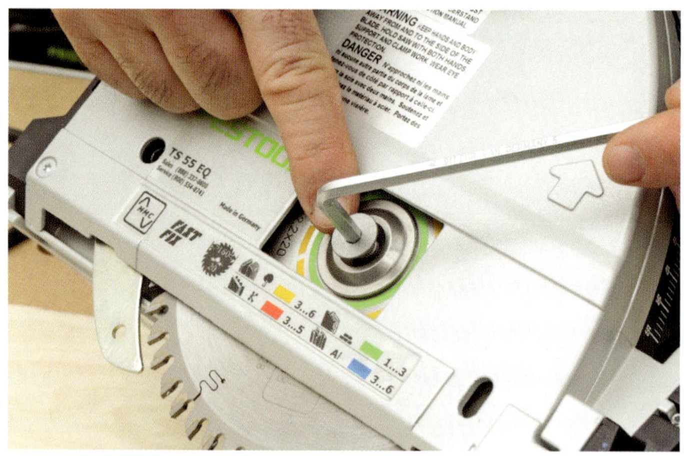

Use the hex key to remove the screw and flange.

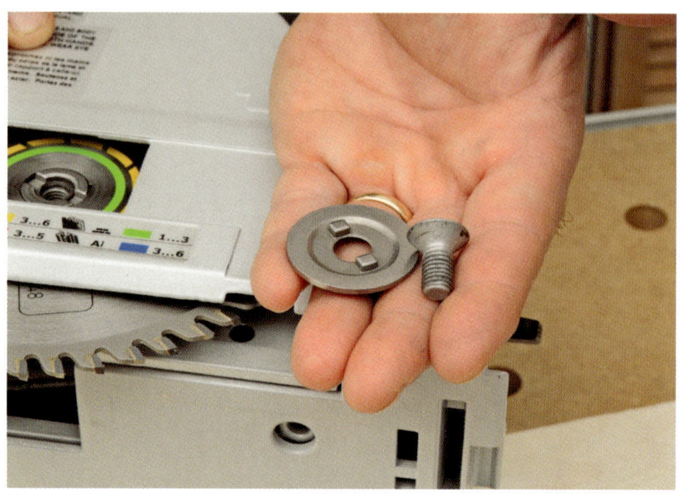

The screw and flange.

Blade Change

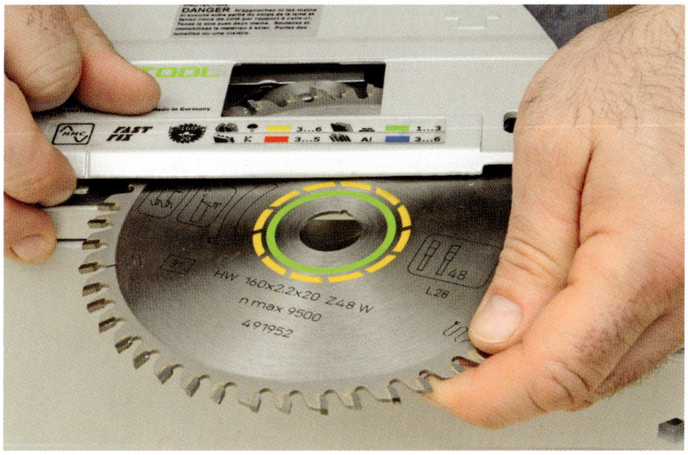

Remove the blade. I lift the kerf knife for convenience.

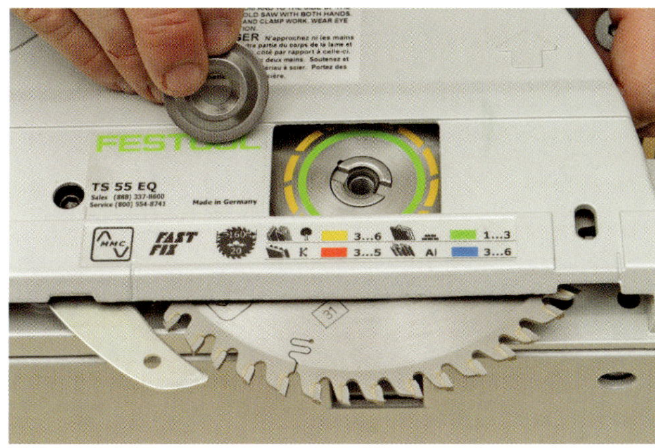

Replace the blade, with the color markings showing through the window.

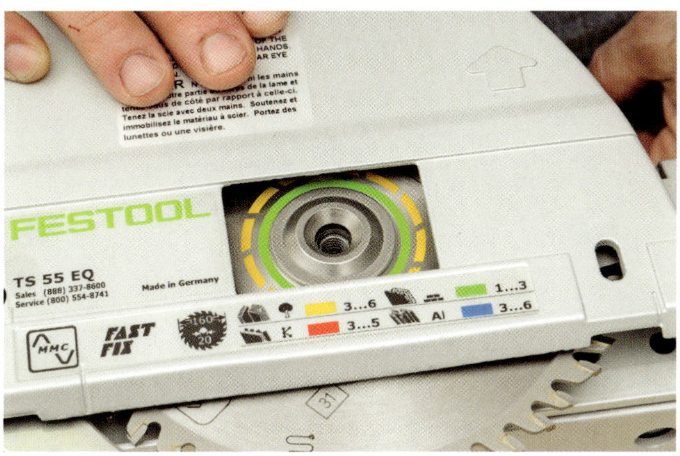

Insert the flange.

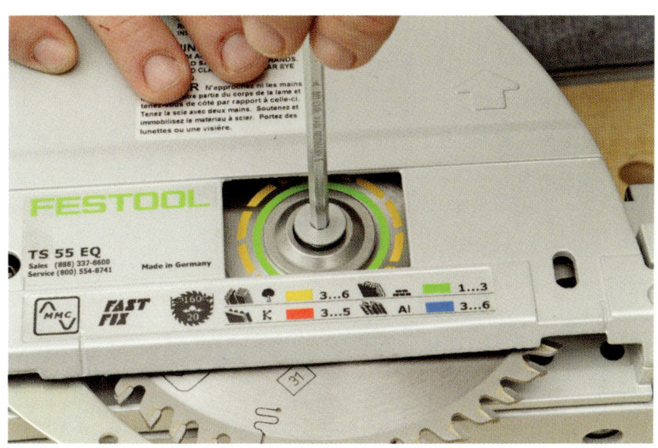

Tighten the screw.

Saw Blades

Model Name		TS 55 EQ (160mm diam.)		TS 75 EQ (210mm diam.)	
	Tooth shape	Model no.	Tooth count	Model no.	Tooth count
Standard saw blade *Tungsten-carbide, saw blade for coarse cuts in wood, building panels and soft plastics.*	ATB	487 377	12	493 197	18
Panther saw blade *Tungsten-carbide, rip blade with radical hook angle and deep gullets for fast, easy rip cuts.*	ATB	495 372	14	493 196	16
Fine tooth saw blade *Tungsten-carbide, saw blade for fine cross cuts in sheet goods, melamine hard and soft woods.*	ATB	495 377	48	495 381	52
Universal saw blade *Tungsten-carbide, saw blade for clean cuts in wood, construction panels and soft plastic. Fine rip blade for glue ready cuts.*	ATB	495 376	28	493 198	36
Aluminum/plastic saw blade *Tungsten-carbide, triple-chip saw blade with negative hook for cutting aluminum and nonferrous metals, acrylics and hard plastics.*	TC	495 373	56	493 201	72
Solid surface/laminate saw blade *Tungsten-carbide, triple-chip saw blade with positive hook specially designed for cutting laminate flooring and solid-surface countertop materials.*	TC	495 375	48	493 200	60
Steel/multilayered board saw blade *Tungsten-carbide, flat-toothed saw blade specially designed for cutting steel cable duts and mild steel up to 1/8".*	TC	N/A	N/A	493 351	36

Saw Blades

The saw blades specifications and sizes are easily found in the accompanying chart. Here is a brief run-down, using the TS 55 EQ blades.

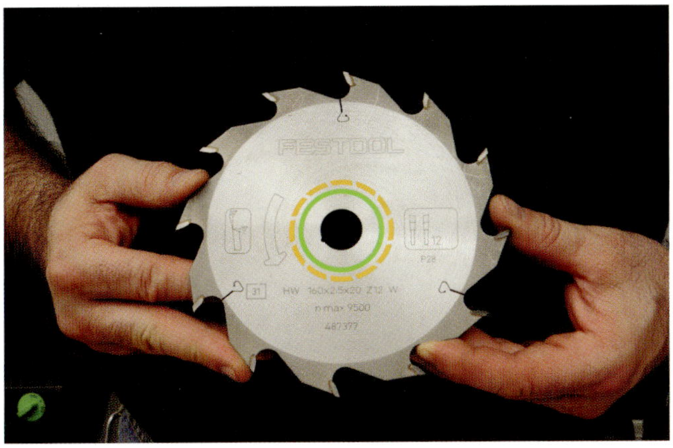

The 12-tooth ATB, 487377 is great for rough carpentry and building panels such as OSB.

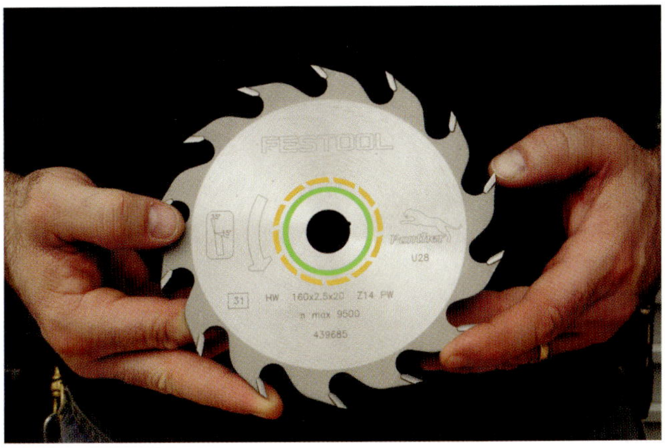

The 14-tooth ATB, Panther saw blade, 439685, is designed for fast, easy rip cuts in solid woods.

The 28-tooth ATB universal saw blade is a good all-around blade for ripping and crosscutting in all woods and plywoods.

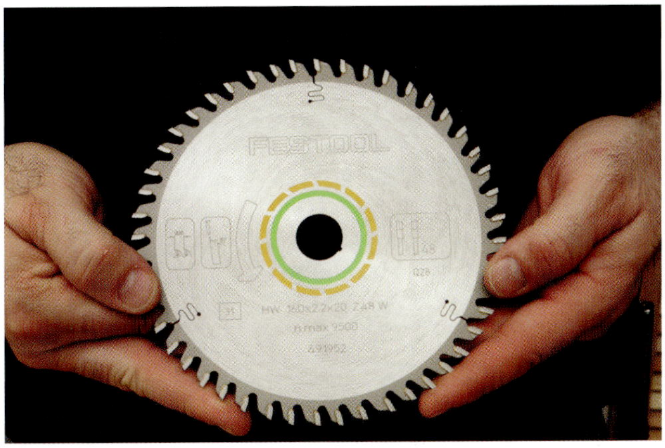

The fine 48-tooth ATB blade comes with the saw. It makes fine cuts in sheet goods, melamine, and hard and soft woods.

Saw Blades

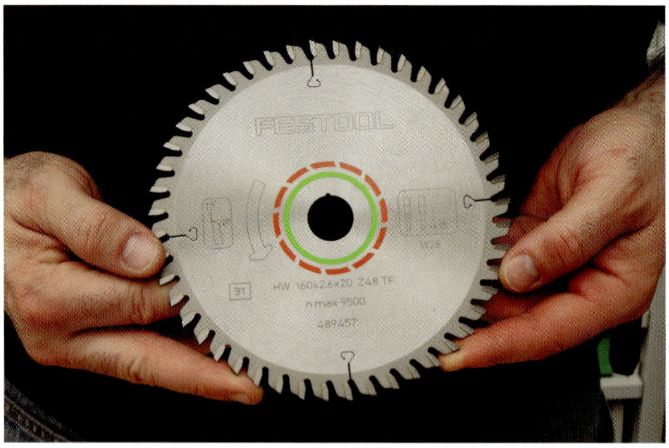

The fine, 48-tooth TCG blade is design for solid surface countertops and laminate flooring.

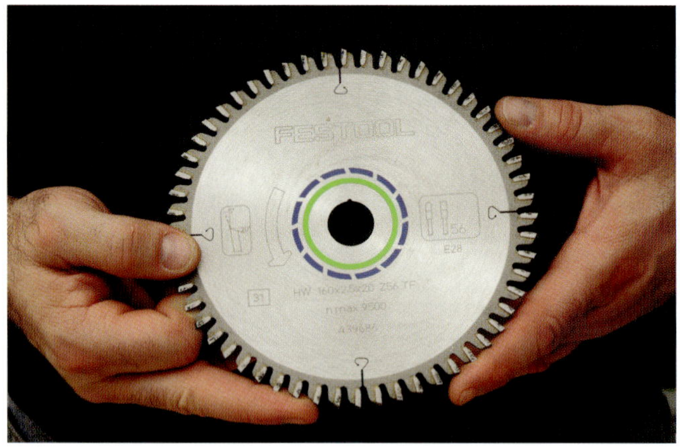

The 56 tooth TCG blade is for aluminum and non-ferrous metals. This is also good for cutting acrylics.

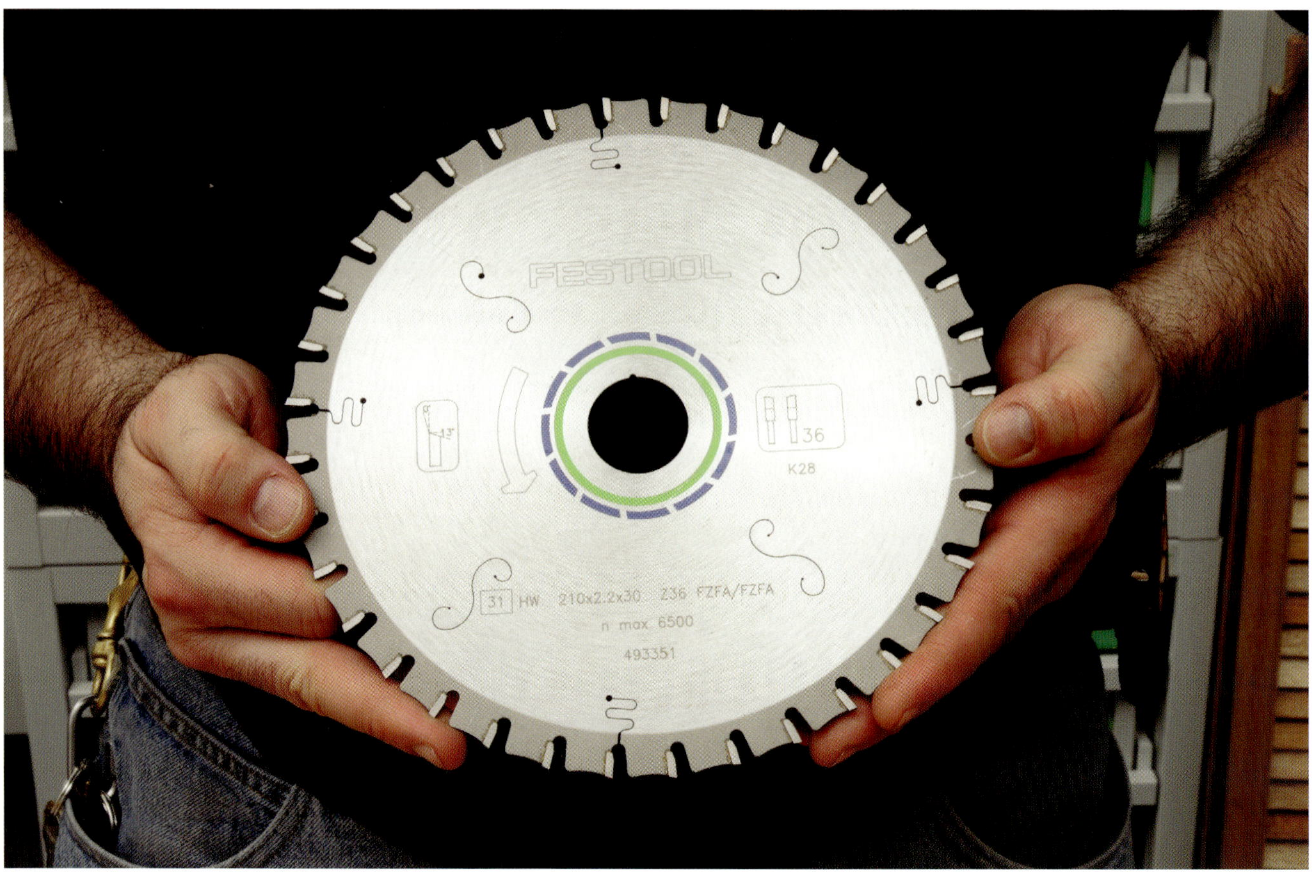

For the TS 75 all the same blades are available (the tooth count is, of course, higher). In addition there is one other blade for the TS 75, unavailable in the smaller size: a 36-tooth blade, specially ground for cutting mild steel up to 1/8" thick. This is great for cutting down steel clad doors.

Variable Speed

The variable speed dial is one of six elements of the Multiple Material Control (MMC) system. The others are automatic and include constant speed under load, soft start, temperature overload protection, and current triggered overload protection.

• **Variable speed** allows the user to set the optimal speed for any material.

• **Constant speed under load** monitors the motor, adjusting the power to varying loads to ensure constant speed. This drastically reduces saw burn.

• **Soft start** makes sure that the motor ramps up its RPM so the saw does not jerk at start-up. This reduces the loss of control caused but sudden spurts of power at start up.

• **Temperature overload protection** shuts down the motor when it overheats due to overexertion. When the motor cools it can be used again.

• **Current trigger overload protection** is essentially a surge protector, shutting off the saw in the event of spikes in current that could damage the electronics or the motor.

The variable speed dial ranges from 1 to 6 ((2000 - 5200 rpm on the TS 55 EQ and 1,350 - 3,550 rpm on the TS 75 EQ). The range is stepless.

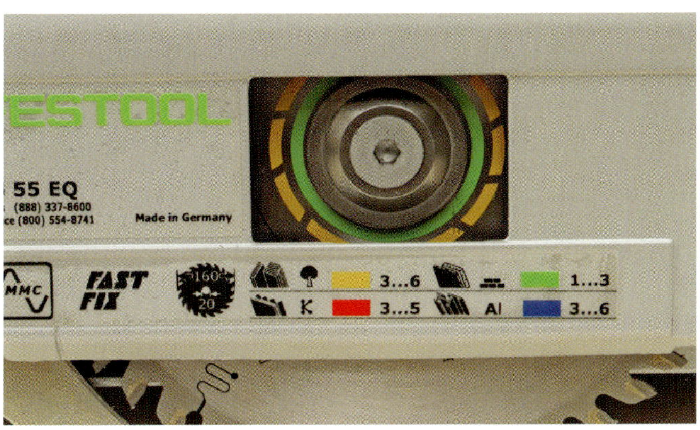

The speed is the only of these features that is manually adjustable. A convenient chart under the blade window provides the code, which references the **color of the broken circle** around the arbor, visible through the window.

• **Yellow=wood.** The speed setting should be 3 to 6. Higher speeds are the norm, but for thicker woods or when there is evidence of some saw burn in the kerf, lower the speed.

• **Red=plastics or solid surface.** The speed setting is 3 to 5, again depending mostly on the thickness of materials. If there is chatter or melting, lower the speed.

• **Blue=aluminum or non-ferrous, softer metals.** The speed setting is 3 to 6. The thicker the material the slower the speed.

• **Green.** Doesn't apply to these machines. The main colors that users should be aware and mindful of when using TS saws are the yellow, read and blue.

Guide Rail Basics

The standard package for both saws includes a guide rail. Various lengths and accessories are available to expand its usefulness.

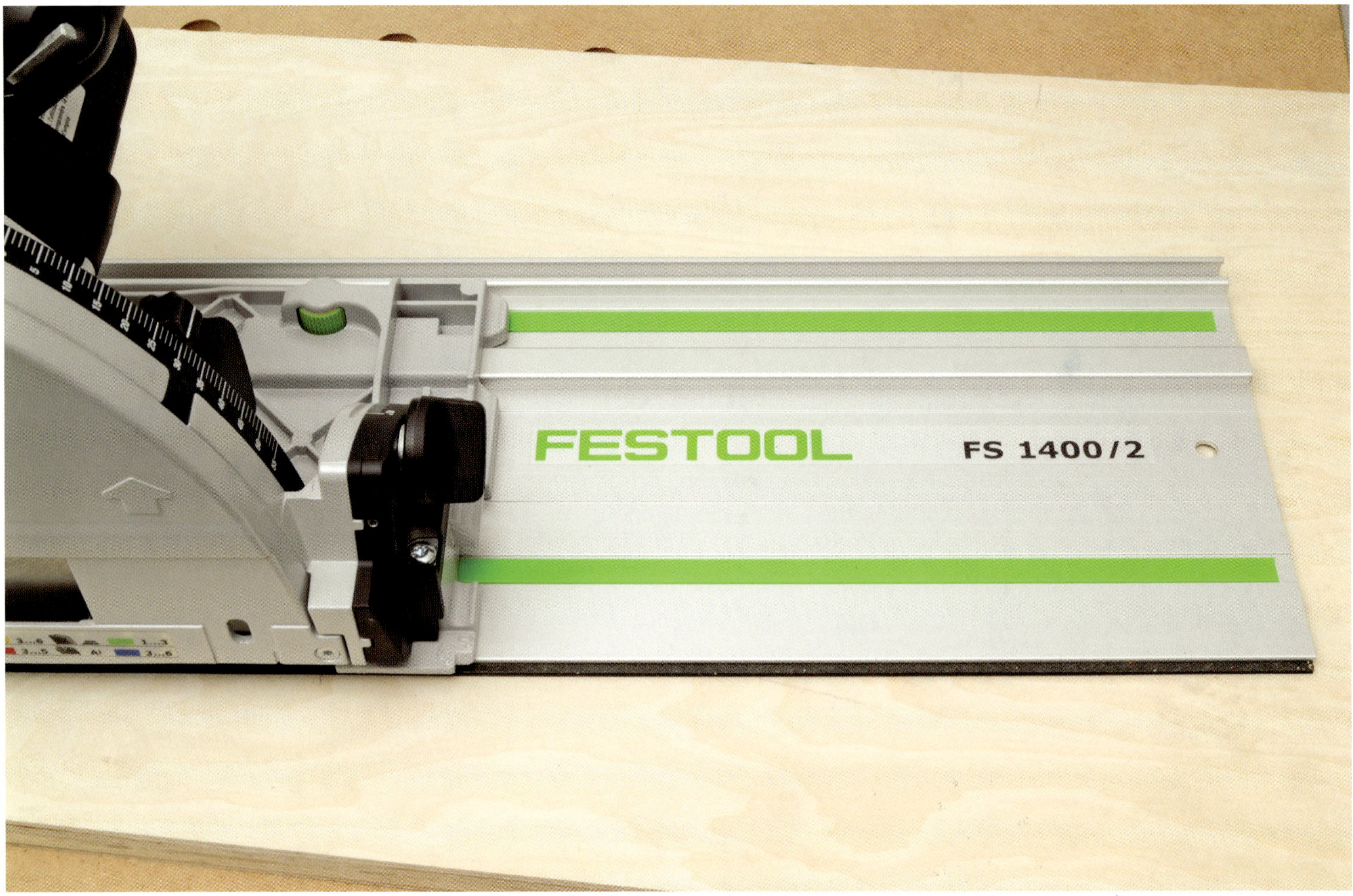

A guide rail is included with both the TS 55 EQ and TS 75 EQ saws. The FS 1400/2 guide rail comes with the TS 55 and the FS 1900/2 comes with the TS 75. The base plate of the saw is designed to run on the extruded channel of the Guide Rail, while utilizing the integral splinterguard, for creating an accurate, straight cut.

Guide Rail Basics

The green wheels on the saw's base plate are used to eliminate any play in the plate as it slides on the guide rail.

Underneath you can see that they move eccentric wheels, increasing or decreasing pressure on the guide rail.

Guide Rail Basics

To adjust, tighten both screws until the saw will not move on the guide rail, then back them off gradually until there is no wiggle, but the saw moves freely on the rail. Any play can be heard and felt if you gently pivot the base plate. Once properly adjusted the saw should have smooth and easy movement on the rail without excess lateral play.

On the cutting edge of the guide rails is a rubber splinter guard. It prevents tear-out or splintering of the material you are working with and allows you to align the guide rail to your scribe or witness marks.

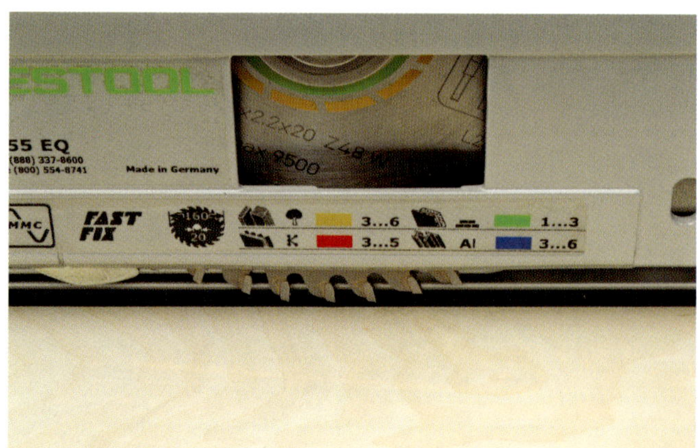

Initially the splinter guard extends beyond the width of the blade. This is designed to be cut by the blade on its first use, giving a precise fit.

Guide Rail Basics

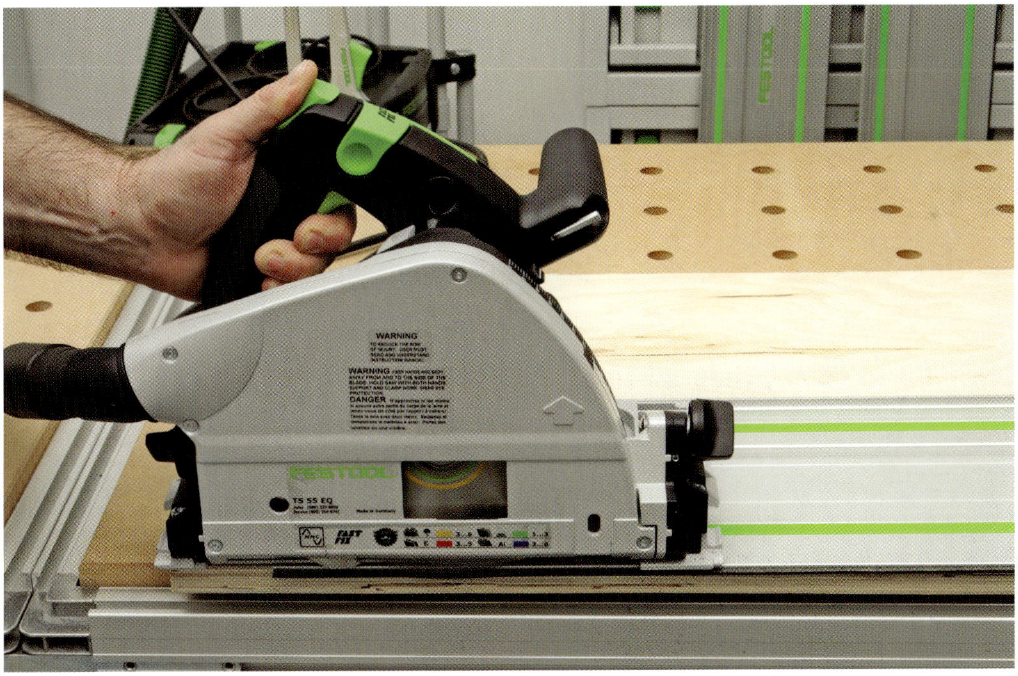

Set the variable speed to 1 and the depth to 10mm. Place the saw on the guide rail and trim the splinter guard. Plunge…

…and cut.

The result is a very snug fit of the blade against the splinter guard. Of course, over time the splinter guard will wear. When you become dissatisfied with how clean your cut is, the guard can be easily replaced.

When that time comes, simply peel back the guard from the rail, remove the glue by scraping or with mineral spirits, and install a self-adhering replacement.

Guide Rail Basics

One of the great features of the saws is thatthey are engineered so the same scribe line is followed with a square cut or beveled cut.

To demonstrate this, a square cut is being made using the guide rail.

Then, without adjusting the guide rail or the position of the wood, a cut of 14 degrees is made.

The cut is stopped so you can see the bevel.

From the top, however, it is clear that the saw followed the same scribe mark with no adjustments other than the angle of the blade.

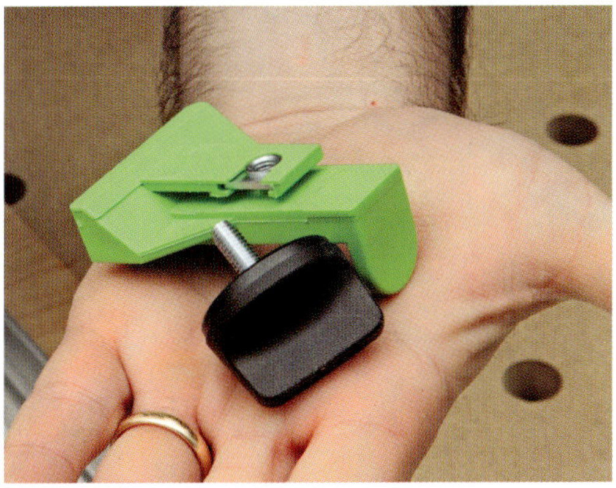

When it is necessary to keep the off-cut splinter free, the saw comes with an off-cut splinter guard accessory.

It clips into the front, lower edge of the blade housing…

…where a knob holds it in place.

Push the tab until the guard is in contact with the wood surface and lock in place.

As the saw blade plunges it cuts the guard...

...creating a zero clearance. With wear it may work less effectively. When you are dissatisfied with the results, the guard can easily be replaced. Replacements are available in packs of five.

Using the splinter guard with the guide rail system results in a splinter-free surface on both sides of the cut.

A Plunge Cut

The saws can make precise plunge cuts. Mark the start and stop points on the scribe line.

Move the guide rail to the line. Note that this splinter guard has had its initial trimming, so it can be aligned with the scribe line for an accurate cut.

A Plunge Cut

The guide rail has slip resistant pads on the underside to keep it in place. For a more secure hold, especially when overhanging a table as here, it is suggested that the guide rail be clamped to the board or table top. Various clamps are available as accessories and are covered later in this book, when the guide rail system is discussed in detail.

At the lowest depth setting for the blade, which is always used for the plunge cut, the blade housing has indicators of blade width at the front…

...and the back.

A Plunge Cut

Align the back indicator with the stop mark.

Move the limit stop into position on the rail at the back of the cut. It goes over the tab in the base plate, keeping the saw from rising during the plunge cut, and acts as a stop at the back of the housing.

A Plunge Cut

Align the front indicator with your mark...

...and abut a second limit stop to the front of the base plate.

A Plunge Cut

With the heel of the plate against the back limit stop, plunge the saw in…

…and cut to the forward limit stop.

The result is a perfectly aligned interior cut.

Installing the Dust Cover Accessory and Parallel Guide

The dust cover accessory, coupled with a 36mm hose, provides even greater dust extraction in particularly sensitive applications, hospitals for example. At the left you see one of two holding brackets, which consist of a clip, a bolt and spring, and a knob.

The holding brackets attach to the base plate of the saw, one at the front and one at the back. A bolt is inserted from beneath and a spring placed over it.

The clip goes over the bolt, bent ends down, and the knob is screwed on.

Installing the Dust Cover Accessory and Parallel Guide

Insert the dust cover rods along the grooves in the base plate. Make it snug to the blade casing at the front and tighten the front knob.

Leave a small gap at the back for air movement and tighten the back knob.

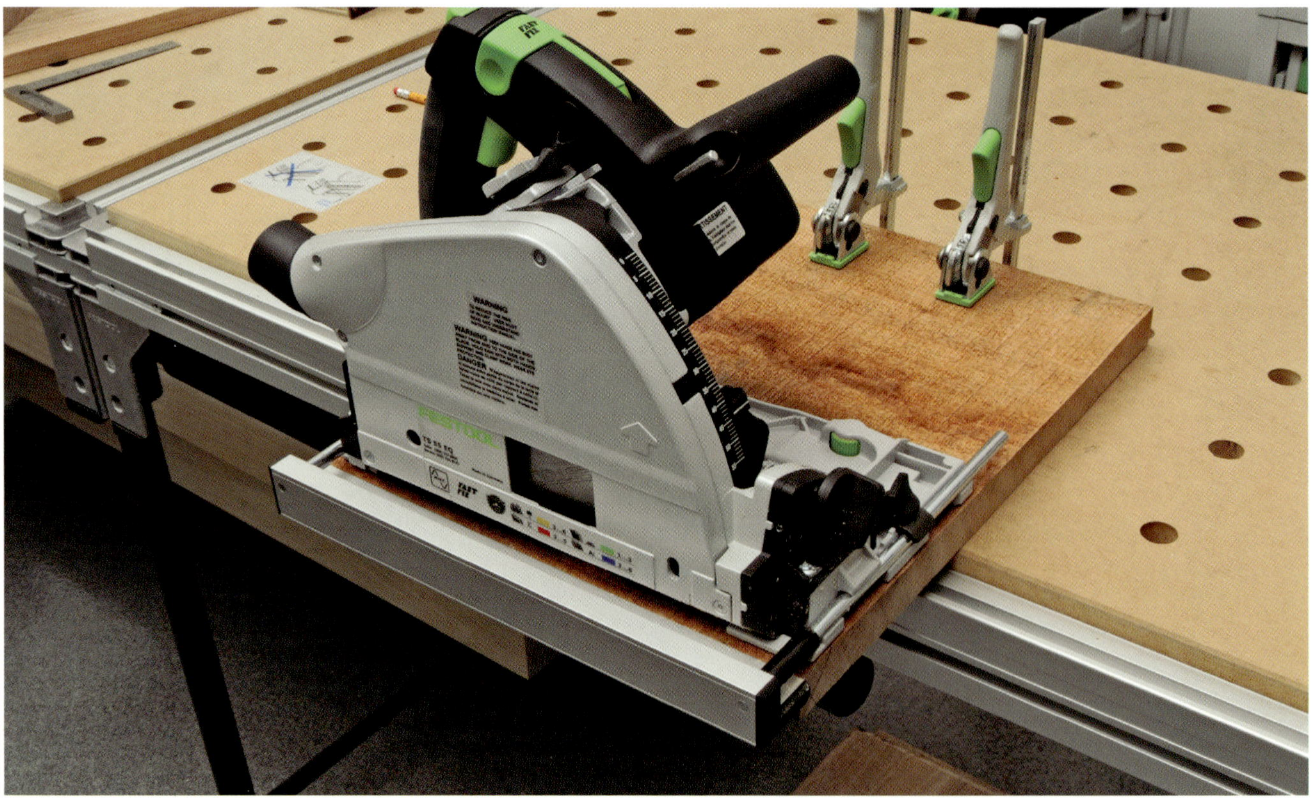

The same clamps are used to install the parallel guide. With proper adjustment, this can be used to make repetitive thin, ripped cuts.

2. THE MULTIFUNCTION TABLE

The Multifunction Table (MFT) provides the solid and versatile workbench that is so important to accurate cutting and assembly work. Designed to be used with the Festool system, it is both sturdy and portable, allowing for reliable use in the workshop and easy transport to the job site. It is the perfect table for cutting, routing, sanding, and more.

The first generation came in two sizes, MFT 800 and MFT 1080. The second generation offers one size for general use, the MFT/3, and one, MFT/3-Kapex, for use with Festool's Kapex sliding compound miter saw. The MFT/3 height has been adjusted to make work more comfortable. It has V-groove channels for faster, easier adjustments, and a redesigned fence and miter gauge.

Multifunction Table features

- Flat and forgiving work surface. The perforated plate is flat and makes a great work surface. It is also replaceable.

- Grid hole pattern. Multiple holes in a uniform pattern are ideal from clamping operations.

- Folding legs make the table easily transportable and also allow for use of the table on the floor.

- Adjustable leg to accommodate uneven surfaces, while keeping the table level.

- Guide rail system for use with TS plunge saws and OF Routers. Use to make cut-offs, angle cuts, compound cuts, and, with routers, dados and flutes.

The first generation of Portable Multifunction Tables came in two sizes, the MFT800 (23" x 28"/590mm x 725mm work area) and the MFT1080 (28" x 45"/725mm x 1165mm work area). In 2008 a new generation was introduced including the MFT/3 and the MFT/3-Kapex. The channels and heights of the new tables have been redesigned, but the interface with the guide rail system and tools is essentially the same. See the side bar on page XXX for information about these new tables.

Multifunction Tables MFT 800, MFT 1080, MFT/3, & MFT/3-Kapex Specifications

Model Name **Item**	**MFT 800** **492 651**	**MFT 1080** **492 652**	**MFT/3** **495 462**	**MFT/3-Kapex** **495 565**
	Discontinued	*Discontinued*		
Standard package includes				
Guide rail supports (2x)	•	•	•	
Angle unit (protractor head)	•	•	•	
Fence rail	•	•	•	
Fence rail clamp	•	•	•	
Adjustable stop	•	•	•	
Deflector	•	•	•	
Guide rail	FS 800	FS 1080	FS 1080	
Specifications				
Table dimensions	23" x 28"	28" x 45"	45-9/16" x 30-7/16"	34-7/32" x 22-7/8"
	590 x 725mm	725 x 1165mm	1157 x 773mm	869 x 581mm
Work space	24" x 20"	42-1/2" x 28"	43-3/8" x 28-1/4"	32-1/16" x 20-11/16"
	625 x 510mm	1085 x 725mm	1102 x 718mm	814 x 526mm
Working height [folded]	32" [8"]	32" [8"]	35-7/16" [7-3/32"]	31-1/8" [7-3/32"]
	820mm [205mm]	820mm [205mm]	900 mm [180mm]	790mm [180mm]
Max. workspace thickness (with FS Guide Rail)	3-1/2" (90mm)	3-1/2" (90mm)	3-1/16" (78mm)	
Weight	44 lbs. (20 kg)	73 lbs. (33 kg)	62 lbs. (28 kg)	42 lbs. (19 kg)

Accessories

Model Name	Model No.	MFT 800	MFT 1080	MFT/3	MFT/3-Kapex
Perforated top *replacement.*	488 565	•			
Perforated top *replacement.*	489 396		•		
Perforated top *replacement.*	495 543			•	
Perforated top *replacement.*	495 544				•
Cross members, *for additional stabilization of the table.*	495 502			•	•
MFT Connector, *for connecting 2 MFT tables (two are needed) Note: MFT/3 does not connect to MFT 1080 or MFT 800.*	484 455	•	•	•	•
Table extension, useful when extra table length or width is needed	486 575	•	•		
Stop flag, *for setting up multiple repeat cuts.*	495 542			•	•
Longitudinal stop, *adjustable.*	488 564	•	•		

2. THE MULTIFUNCTION TABLE

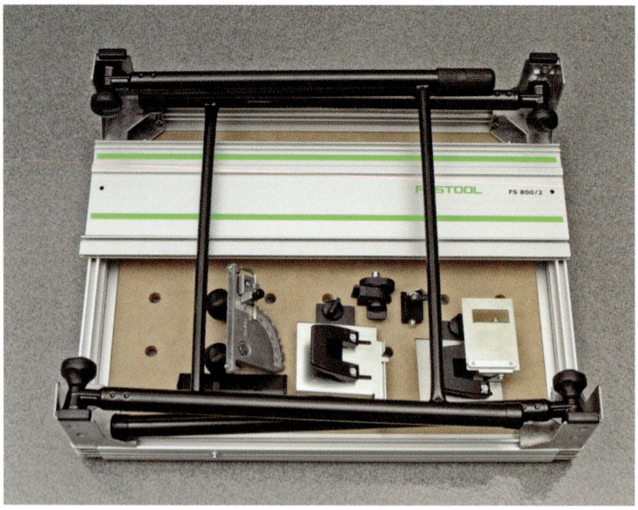

Usually the MFTs are purchased with the hardware kit as shown here, though the table and the kits can be purchased separately. See the Specifications on page 38 for details about the accessories included in the hardware kit.

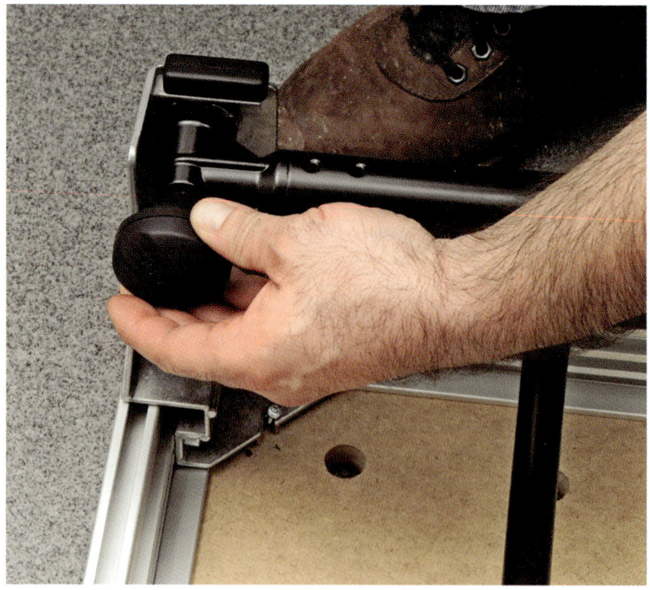

To open the legs, begin by loosening the leg nuts.

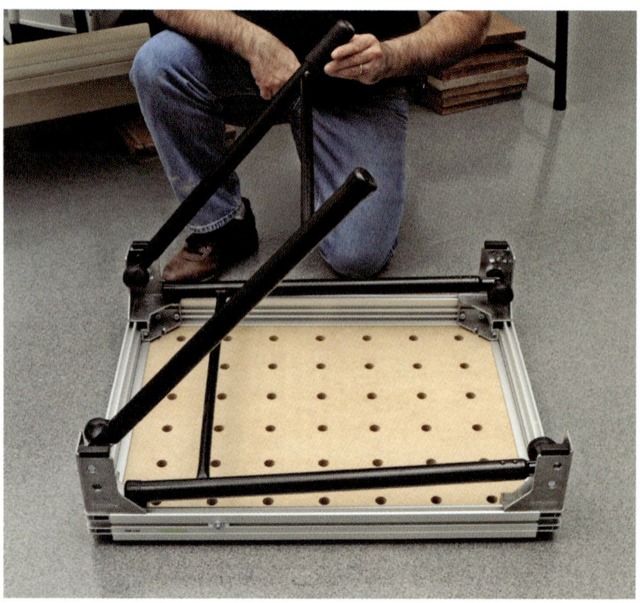

Open the leg until it snaps in place.

Tighten the leg nuts.

Some applications call for a low station. The corner brackets have rubber pads so that when the legs are closed...

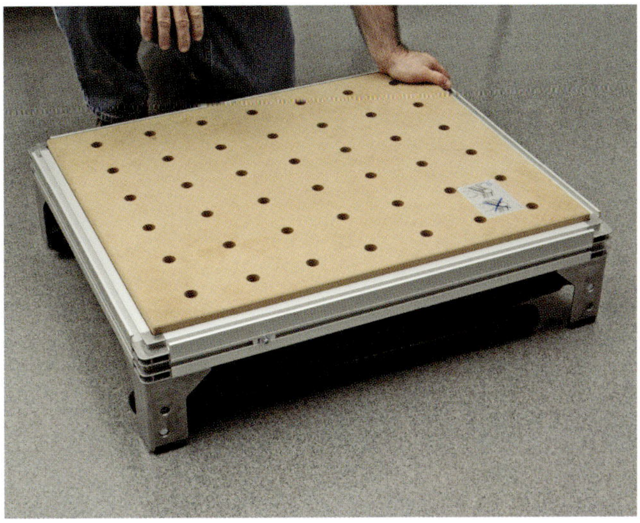

...they provide short legs.

For uneven surfaces one leg is adjustable. Adjust the length until the wobble is eliminated.

Ready to go.

2. THE MULTIFUNCTION TABLE

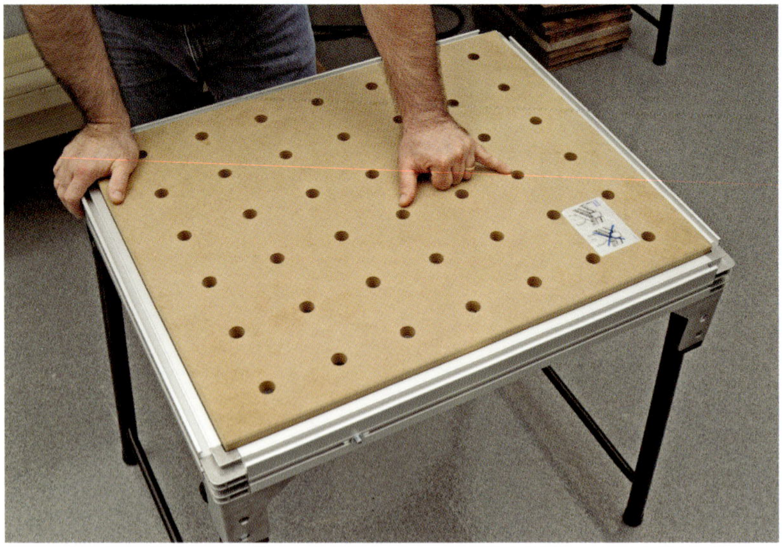

The MDF work surface is perforated with an array of 20mm holes, precisely laid out for accuracy.

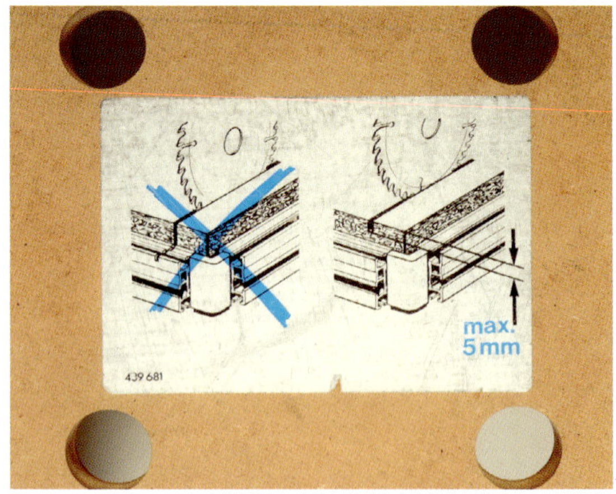

Please note that the maximum depth of the cut into the MDF work surface is 5mm. Going deeper risks cutting into the structure of the table and damaging both it and the saw blade.

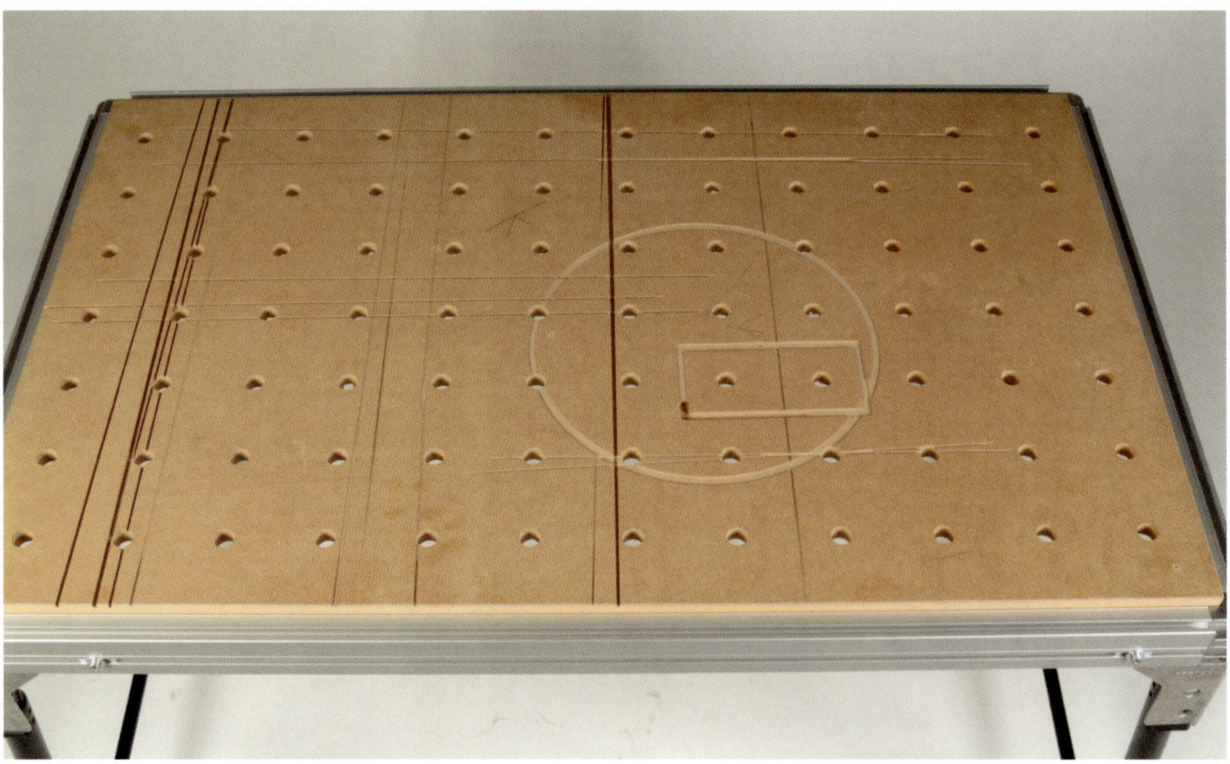

As beautiful as it is, this is a replaceable support surface. It acts as a spoil board and is MEANT TO BE CUT INTO! When one side finally becomes too ragged to use, the top can be turned over for a fresh surface. And when that side has seen the end of its usefulness, a new surface can be ordered and is easily installed.

Clamps

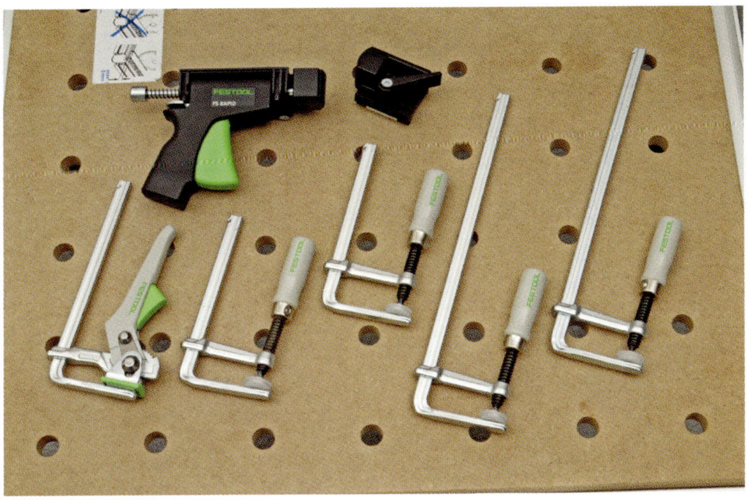

Four clamps that work with the guide rails include the FS-Rapid clamp and fixed jaw (top), the ratcheting Quick Clamp, and the 120mm and 300 mm screwing clamps that are packaged as pairs.

The basic clamp can be used in several ways. To secure the wood to the work surface, run the head through the table and tighten down…

…run the head up from the bottom and tighten from below…

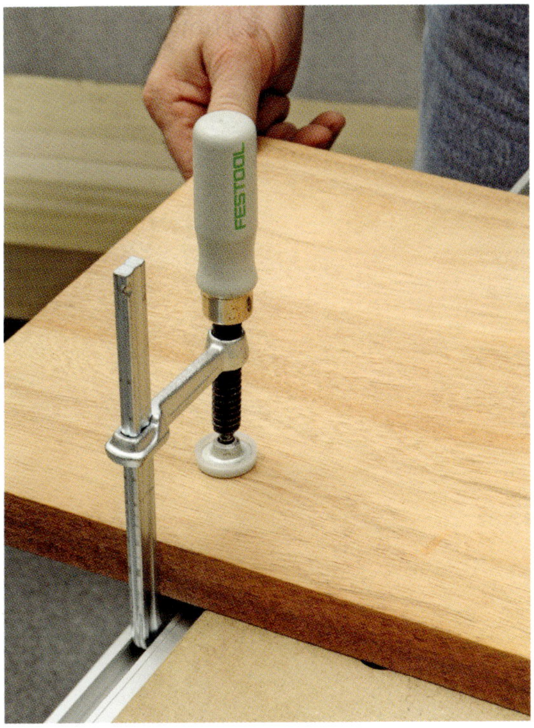

…or run the head in the top channel and tighten down.

Clamps

You can also use two clamps in the side channels to hold the wood vertically.

The longer clamp can be used for deeper objects.

The rapid clamp can be used for the same functions, and lock down by simply lifting the handle. It is released with the green trigger.

The FS-Rapid clamp is versatile. Here it is used on a Clamping Profile. Inserted into the groove and tightened, the tail piece holds the piece and the green trigger in the head piece tightens the grip. This clamp will also fit in the grooves of the guide rail.

Clamps

Two clamping elements are packaged together. They consist of a tail piece (A) and a head piece (B) and lock knobs (C) to secure them to the table from beneath when they are needed. The tail and head pieces fit into the holes of the work surface.

The tension of the headpiece is adjustable with a 3mm hex key inserted into this socket.

Clamps

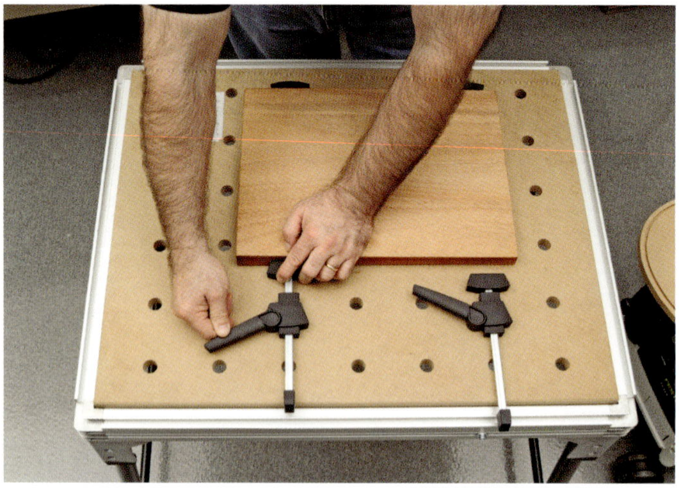

With the wood against the tailpieces adjust the headpieces with the levers to secure it.

The clamping system in place.

The round edge of the dogs and the table design permit this clamping system to be used with any number of shapes.

Joining Tables

Two tables can be joined using the MFT connector.

When joining end to end, fold up one set of legs where the tables will meet. The adjustable legs should be on the outside.

Secure the connector in the slot of one table, tightening the screws.

Joining Tables

Join the tables.

Tighten the remaining screws.

Multiple tables can be joined in this way. Note here that the tables are being joined end to side, so the leg is not folded up. Also, this requires only one connector, on the outside edge.

Table Accessories

The table extension provides about two feet of extra support for the work project.

From the bottom side you can see the attaching bracket, with a key that inserts into the side channel of the table.

This adjustable positive stop is set after leveling, allowing for accurate repositioning of the extension. The outside hex screws are used for adjustment.

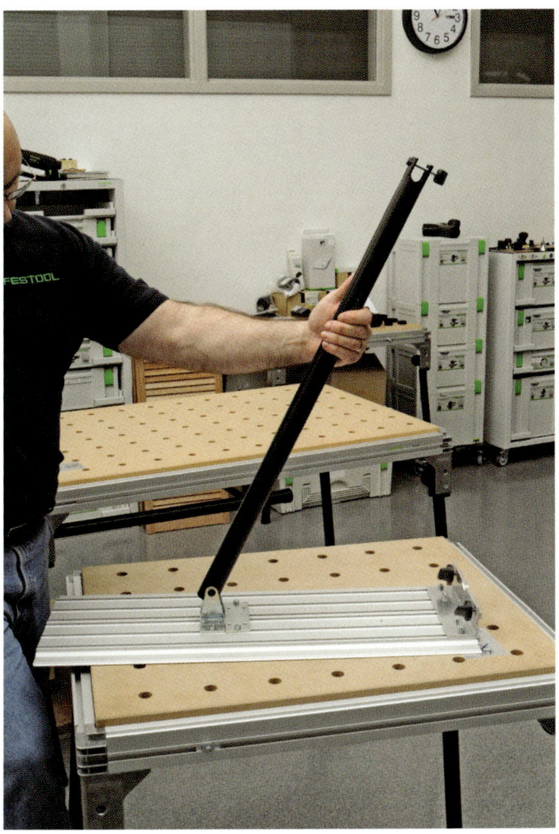

The adjustable arm bracket clamps to the stretcher between the legs.

An adjustable arm bracket for leveling. Again the outside hex screws are used for adjustment.

Slide the key into the channel of the table.

Attach the arm bracket to the stretcher.

Using a straight edge for a guide, at the connecting bracket adjust the height of the extension to match the table. Lock in place.

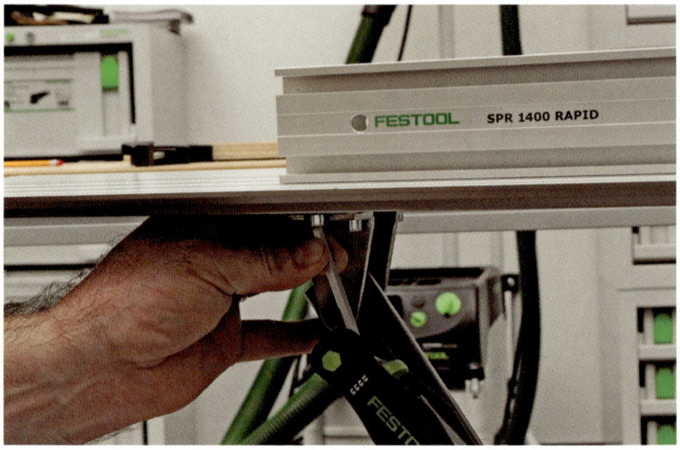

Adjust the leg bracket so the extension is level. Lock in place with the hex screws.

Table Accessories

Slide the positive stop so it is snug against the leg bracket. Lock in place.

Complete

Set-Up: The Hardware Kit

There are two sides to the table, front and back. On the front side, the factory-set stop is on the right, on the back it is on the left. In this view the back is toward the camera.

On both supports, the black key on the base plate goes in the side channel and the metal hook rides in the top channel on the MFT 800 and MFT 1080. For the MFT/3 see the sidebar.

The guide rail support has two parts, a front and back. The back support, shown here, has a hinged plate, the front does not.

Insert the hinged support in the side channel at the back of the table, and slide to the stop. Lock in place using the knob.

Set-Up: The Hardware Kit

Pull the hinge plate all the way up and lock in place with the lever.

The front support slides in from the left to the stop. Lock in place with the knob. Pull up the metal plate and secure it with the black lever. Note the locating pin on the top edge of the plate.

The key on the face of the hinge plate is inserted into the channel of the guide rail.

Set-Up: The Hardware Kit

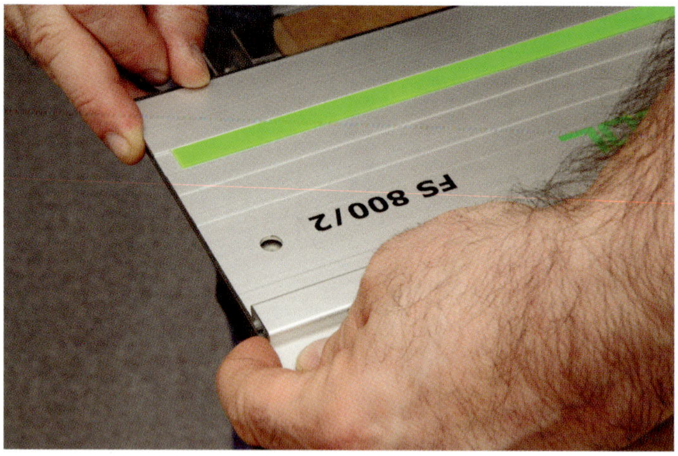

On the MFT 800, adjust the guide rail until its edge is flush with the hinge plate.

On the MFT 1080, the edge of the hinge plate should fill about 1/3 of the hole at the end of the guide rail. This placement maximizes the capacity of the TS saws on the guide rail.

The channel fits over the locating pin on the front, non-hinged support.

Insert the short arm of the hex key in the outer most screw of the hinge plate and tighten. This gives added torque.

Set-Up: The Hardware Kit

Lift the guide rail and insert the long arm of the hex key into the other screw and tighten with lighter torque.

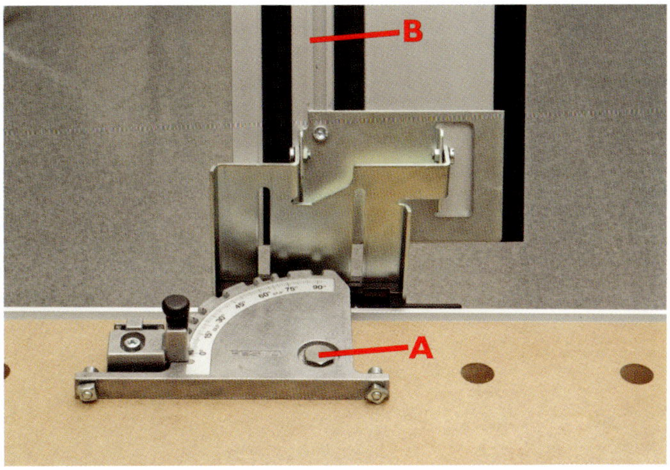

With the guide rail at the factory stop position, install the angle unit or protractor head so the large nut (A) is in alignment with the channel (B) of the rail. Lock it in place using knobs underneath the table.

The square nuts on the edge run into the channel of the fence. The black knob sets the angle of the fence. Here the fence is being installed in its lower, flat position.

It can also be set upright.

Set-Up: The Hardware Kit

The fence should extend about half way or less into this hole. This is an optimal setting and prevents cutting into the fence when doing bevel cuts.

Tighten the fence using the hex key.

The angle stop slides into the upper channel of the MFT 800 and MFT 1080 (see sidebar for the MFT/3). It rests inside the T-slot in the fence to stabilize and secure it. Push the positive detent pin when the protractor is at the desired angle, while pressing the clamp against the fence rail and engaging.

The adjustable stop can be set on the fence for repetitive cuts.

Set-Up: The Hardware Kit

The last piece in the Hardware Kits for the MFT 800 and MFT 1080 is the deflector, which goes on the end of the rail and keeps the dust extraction hose and cord from catching on the rail.

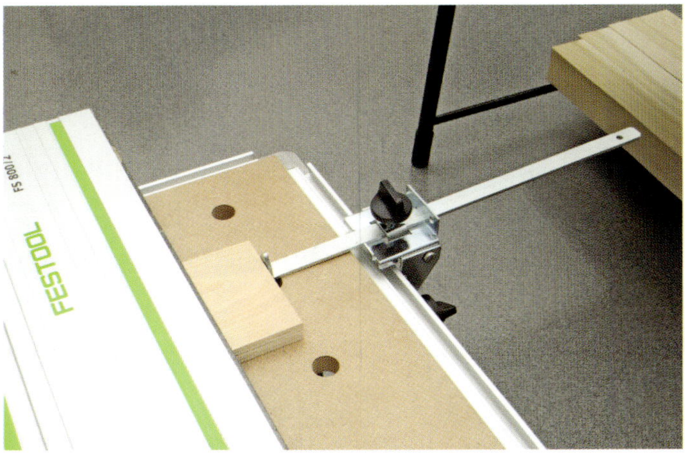

The longitudinal stop is an optional accessory and is used for pieces less than 7" wide when you are making repetitive cuts. It attaches to the side channel of the outboard end of the table. Set it to the length desired remembering to allow for the kerf.

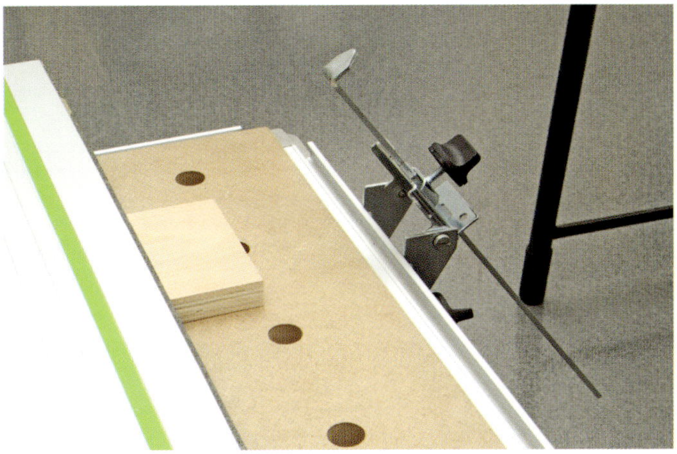

When the length is set pivot the stop out of the way. Very important!

The MFT/3

In 2008 Festool introduced the MFT/3 multifunction table as the latest evolution in its work platform. It is taller than the MFT 800 and MFT 1080, ergonomically designed to be a comfortable and safe working height.

While the guide rail hardware is essentially the same as in the earlier MFT work tables, the angle unit and fence have been redesigned to be more versatile and accurate.

The MFT/3 continues the great features of the its predecessors. The legs fold for easy transport or for use on the ground, and the tip of one leg is adjustable for leveling.

The MFT/3

The perforated, reversible work surface is also continued. Although the table comes only in one size, it is slightly larger than the MFT 1080, measuring 45-9/16" x 30-7/16" (1157 x 773mm). Despite its larger work surface, the table is actually 15% lighter than the MFT 1080. The MFT/3-Kapex is shorter and has a smaller work surface, 34-7/32" x 22-7/8" (869 x 581mm). It was specifically designed as a support for the Kapex sliding compound miter saw.

On the MFT/3 the table is connected with a threaded bolt. This insures that the perforations will be square with the table frame and, thus, with the guide rail and other accessories.

One of the main design advances are the V-groove channels on the apron, which make adjustments both faster and easier.

As with the MFT 800 and MFT 1080, the black key of guide rail support slides into the channel on the side of the table with the metal hook riding in the top V-groove channel.

The MFT/3

The hinged support is on the back side of the table and slides into place against the factory installed stop. Remember on the back of the table this stop is toward the left end. Slide the support into place from the right and lock with the knob. Raise the hinge plate to its full height, using the lever to secure it.

The front support slides in from the left to the stop, is locked into place, and the metal plate raised and secured. As before, this support has a metal locating pin to fit into the guide rail.

Install the guide rail, sliding it over the key of the hinge plate on the hinged guide rail support.

As with the MFT 1080, the edge of the hinge plate should fill about 1/3 of the hole at the center of the guide rail. This placement maximizes the capacity of the TS saws on the guide rail.

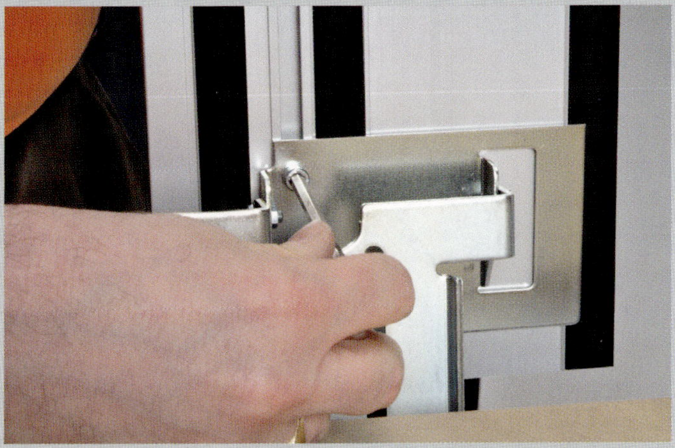

Use the two hex screws under the hinge plate to tighten the guide rail.

The channel fits over the locating pin on the front, non-hinged support.

The miter gauge and fence have been redesigned for greater flexibility and accuracy. The gauge now covers 180° and is movable, connected to the table with a bracket.

Mount the bracket so the V-shaped key runs in the top channel. Running in the space below the V-channel is a clamp that tightens to secure the bracket to the table.

The large knob under the bracket tightens the clamp. With this bracket design, the miter gauge can easily and quickly be moved to any position on the table or removed completely if desired.

The miter gauge is connected to the bracket with an adjustable arm, adding to the flexibility.

The knob on the side locks it in position.

Notice that the back edge of the fence, where it connects with the miter gauge, is notched out.

The keys on the face of the miter gauge run into the channel of the fence. In the flat position, lay the fence on the table and begin to slide the notched edge under the key.

The keys slide into the channel.

Continue until the edge of the notch is flush with edge of the miter gauge.

The forward knob on top of the miter gauge locks the fence in position.

In its upright position than the fence is taller than on previous models.

If adjustment for square is ever needed, loosen the two hex screws behind the postive detent pin, adjust the setting, and retighten.

The miter gauge has preset slots for the most commonly used angles.

To set the angle, simply loosen the protractor knob…

…lift the positive detent pin…

…and rotate the fence until you are at the desired angle and the pin pops into place.

Tighten the knob.

To set the miter gauge for an unslotted angle, loosen the knob, lift the positive detent button, and give it a quarter turn so it rests on the shoulder of the housing.

Select the desired angle...

...and lock the protractor head in place.

Squaring the Rail to the Fence

With the angle unit set to 0°, check that the guide rail is square to the fence.

If adjustment is needed, loosen the angle stop.

Loosen the set screw at the protractor, freeing the fence.

With the protractor set at zero degrees, depress the positive detent pin.

Squaring the Rail to the Fence

While holding the pin, adjust the fence until perfectly square with the rail...

...tighten the fence rail clamp...

...and the set screw in the protractor

Adjusting the Height of the Guide Rail

The guide rail adjusts for various thicknesses of wood. With narrow boards, release the hinged guide rail support, press down on the rail at the board, and lock the hinge support in place.

Move the other end of the rail, release the support, and, while pushing down on the rail over the board, locate the channel over the pin and adjust the support so the guide rail touches the wood.

Adjusting the Height of the Guide Rail

With a wider board the process is similar, but both supports can be loosened at the same time and adjusted.

The router in use. Note that the limit stop on the rail is used for an accurate end point and repetitive work.

With the rail adjusted it is ready for use with a variety of tools.

Angled Cuts Using the Protractor Head and the Multifunction Table

Measured angle cuts can be made using the angle unit with the guide rail, as seen earlier, or by using the protractor head angle unit usually included with the Multifunction Tables (MFT).

For the square cut the protractor was set at 0 degrees.

Loosen the rail clamp…

...and move the fence to the desire angle, in this case 22.5 degrees.

Angled Cuts Using the Protractor Head and the Multifunction Table

Depress the positive detent pin on the protractor.

Lock the fence rail clamp.

Load the board and align your scribe mark with the guide rail.

Angled Cuts Using the Protractor Head and the Multifunction Table

With the saw on the rail, set the depth of the cut (thickness of the wood + 5mm +2mm) and cut the angle.

The result.

Compound Bevel Cuts Using the Protractor Head and the Multifunction Table

The compound bevel combines the angle of the cut and the angle of thc blade. Using the same 22.5 degree for this example, the angle of the saw can also beveled.

Set the depth gauge. At a 30 degree angle or less, the same formula as a straight cut can be used, the thickness of wood + 5mm for the guide rail + 2mm for cut through. Above 30 degrees put the saw on the guide rail on the material, set the saw to the desired angle.

Plunge the blade until it hits the table.

Compound Bevel Cuts Using the Protractor Head and the Multifunction Table

Move the depth gauge up to the stop.

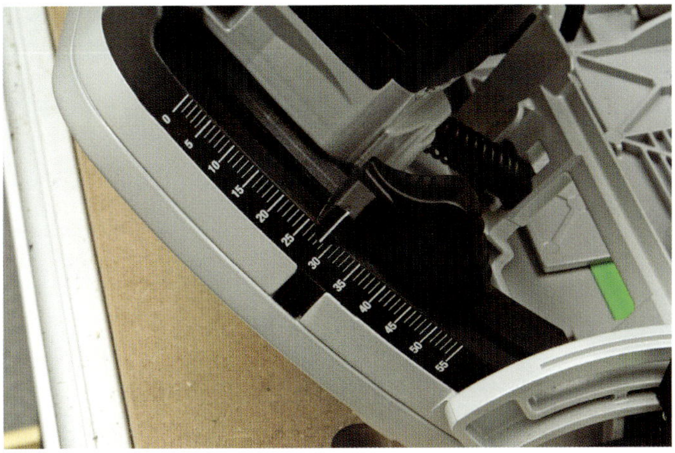

Add 1 or 2mm of depth.

For this compound bevel cut, the bevel of the cut will 22.5 degrees.

The result.

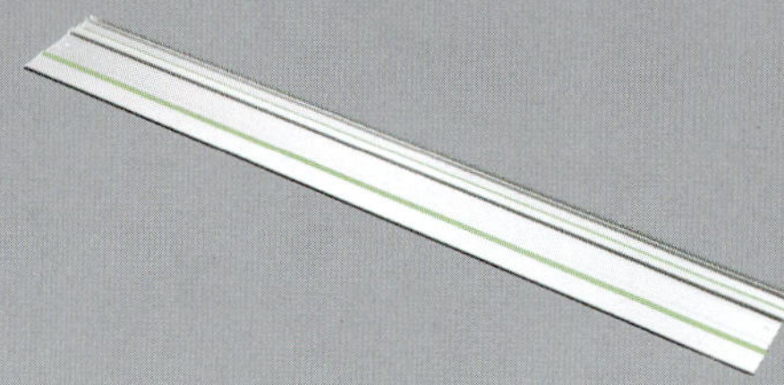

3. THE GUIDE RAIL SYSTEM

The TS plunge saws, and other Festool tools, work hand-in-hand with the guide rail system. Invented over 40 years ago, again and again the guide rail system has proven itself to be an essential element in accurate, efficient handling of these premium tools.

The rails come in eight sizes ranging from 32" (800mm) to 197" (5000mm). The rails join together to create almost limitless possibilities for nearly every work situation.

In addition there are rails in 42" (1080mm) and 95" (2424mm) that are used with the LR 32 hole-drilling system, making it easy to produce rows of holes for cabinets, casework, furniture, and shelf supports. The guide rails are used to precisely index the position of the router for creating the holes, and can also be used to created recessed for hardware like hinges.

The rails can be used in conjunction with the MFT tables or on their own. Soft, non-slip strips on the underside prevent damage to the work piece, and hold the guide rail securely in place.

The rails are equipped with a patented splinter guard, as described earlier. The splinter guard is cut by the blade on the first pass so it fits snuggly against the blade. Once cut, it represents the actual cut line of the saw so that it serves as a reference that can be pressed against the scribe line for an accurate cut. Due to this design a full scribe line is no longer needed. Simply mark the start and stop points, align the guide rail by the splinter guard to those points, and make the cut.

The saw and the guide rail work together, so the saw slides smoothly, producing a precise cutting tool that will remain constant over years of use.

Guide Rail features

- Zero clearance guide rail ensures that the cut mark matches the cut line.
- Quick set-up with non-slipping bottom strips
- Limit stops attached to the rail allow for repeat cuts.
- Effortless guidance of plunge saws or routers
- As part of an integrated system, it works with the Festool TS saws, OF routers, and PS jigsaws.
- Guide strips reduce surface friction with the saw base plate.
- Portable, lightweight, and precise.
- Connectors allow the creation of custom lengths.

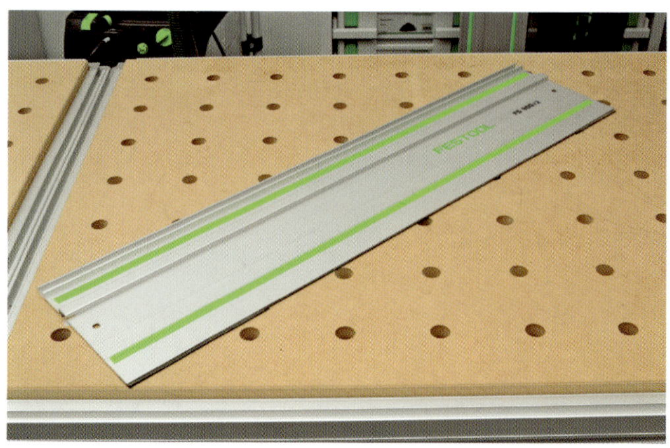

The heart of the system is the guide rail itself. Precision engineered of extruded aluminum, it comes in a 8 lengths ranging from 32 inches (800mm) to 197 inches (5000mm). Two others are available that are indexed to work with LR 32 hole drilling set for making shelf pin holes, Euro hinges and 32mm hole space boring.

Specifications

Model Name Item	FS 800/2 491 499	FS 1080/2 491 504	FS 1400/2 491 498	FS1900/2 491 503	FS2400/2 491 502	FS2700/2 491 937	FS3000/2 491 501	FS5000/2 491 500	FS1900/2-LR 32 491 6212	FS2424/2-LR 32 491 622
Length inches* Length mm*	32" 800mm	42" 1080mm	55" 1400mm	75" 1900mm	95" 2400mm	106" 2700mm	118" 3000mm	197" 5000mm	42" 1080mm	95" 2424mm

* **Note:** The metric dimensions are binding or absolute. Imperial measurements are coarse conversions.

Accessories

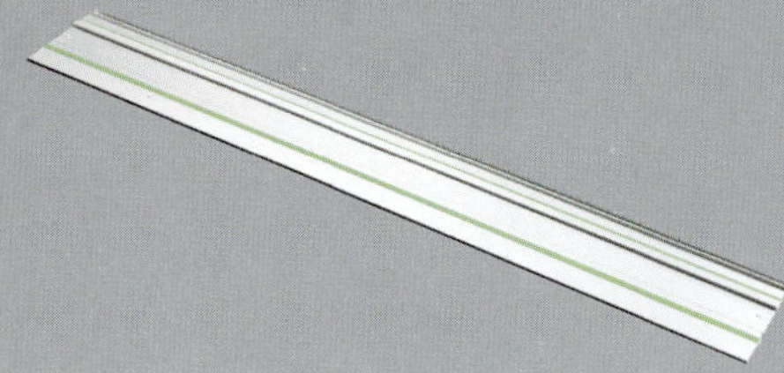

Model Name	Model No.
Angle unit, *for setting angles with the FS guide rails*	491 588
Deflector, *prevents suction hoses and cords from catching on the end of the guide rail.*	489 022
Limit stop, *provides front and rear stop position on guide rail*	485 827
Guide rail tote bag, *for up to two FS1400/2 or shorter guide rails.*	495 544
Gecko.	495 502
MFT Connector, *for connecting 2 MFT tables (two are needed) Note: MFT/3 does not connect to MFT 1080 or MFT 800.*	484 455
Table extension, useful when extra table length or width is needed	486 575
Stop flag, *for setting up multiple repeat cuts.*	495 542
Longitudinal stop, *adjustable.*	488 564
Clamps	
FS-Rapid *clamp with fixed jaw, for clamping and positioning with Festool guide rails— includes fixed jaw and clamping jaw.*	489 790
Clamping profile, *for use with the FS-Rapid clamp to make a bar clamp. Length 32" (800mm)*	490 189
Quick clamp, *easily lock and quickly released for use with MFT and guide rails.*	491 594
Screw clamp, for securing guide rails.	
Clamping thickness 4 11/16" (120mm).	489 570
Clamping thickness 11 3/16" (300mm).	489 571
Clamping elements, for safe and exact fitting of workpiece	488 030

The guide rail assortment.

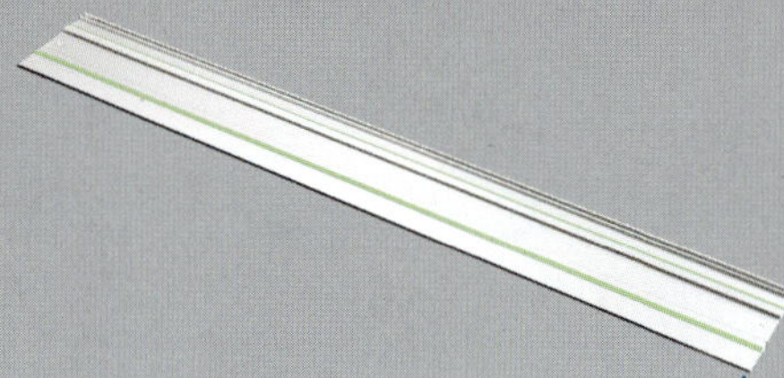

3. THE GUIDE RAIL SYSTEM

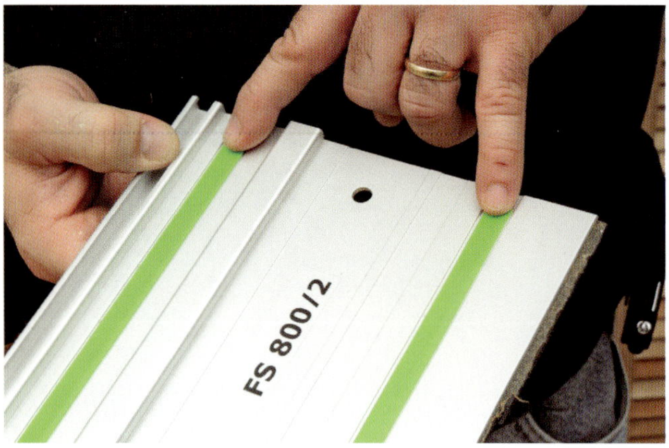

The surface has two low friction glide strips.

This extrusion is the guide for TS saws and the OF routers.
The channel below is for clamps and connectors.

This channel is used for clamping, connectors, and as a tool guide for the PS jigsaw and other tools.

The replaceable splinter guard runs along one edge.

Two foam rubber strips hold the guide in place on the work surface, and are effective enough to obviate the need for clamping in many applications.

While it is always recommended that the guide rail be clamped in place, in some situations this is impossible. For quick and easy cuts, the rubber strips on the underside of the guide rail keep the guide rail from slipping. For critical and finish cut on expensive materials clamps should always be used.

As they wear, the friction glide strips, splinter guards and foam rubber gripper strips are easily replaced with self-adhesive strips.

Guide Rail Accessories

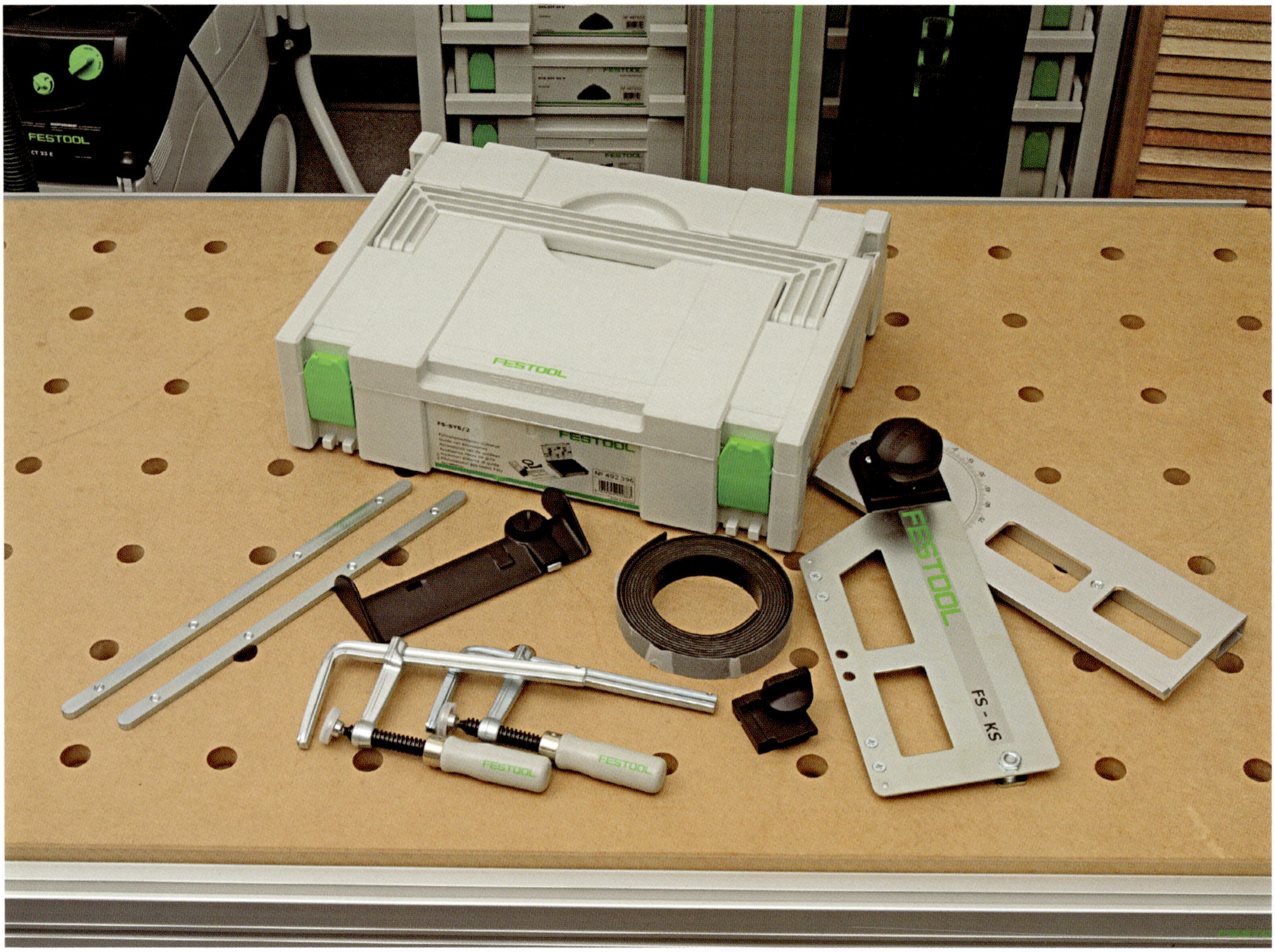

The optional Guide Rail Accessory Kit comes in a SYSTAINER and has useful accessories for the system. Left to right: 2 guide rail connectors, 2 clamps, a hose and power cord deflector, replacement splinter guard, 1 limit stop, and 1 angle unit.

CONNECTORS

Guide rails are designed to be joined together to make longer pieces. This is done with connectors that fit into the channels. The connectors are fitted with slotted screws for fast and easy tensioning when joining rails.

To make the connection insert the connectors into one piece of guide rail, about half the length of the connector, and tighten. One connector tightens from the top, the other from the underside.

Insert the connectors into the other guide rail segment.

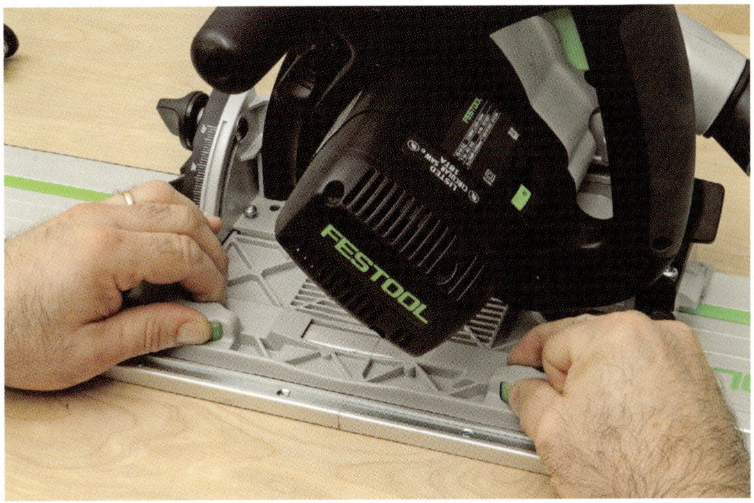

For fine tuning, a simple and accurate alignment can be made by bridging the joint with the saw on the guide rail and tightening the eccentric cam knobs.

Tighten the exposed screws in the connector, then remove the saw, and tighten the screws beneath.

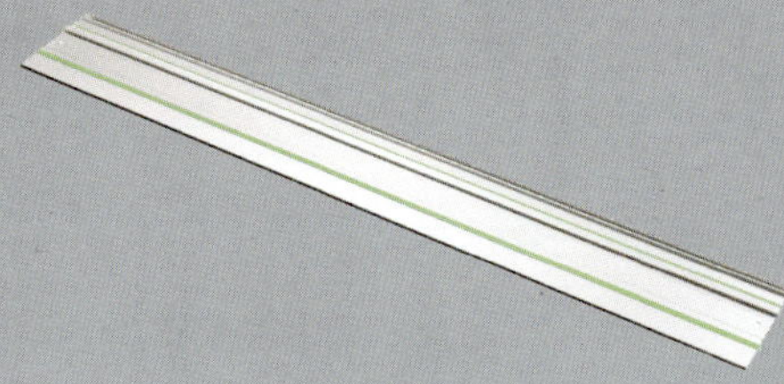

THE ANGLE UNIT

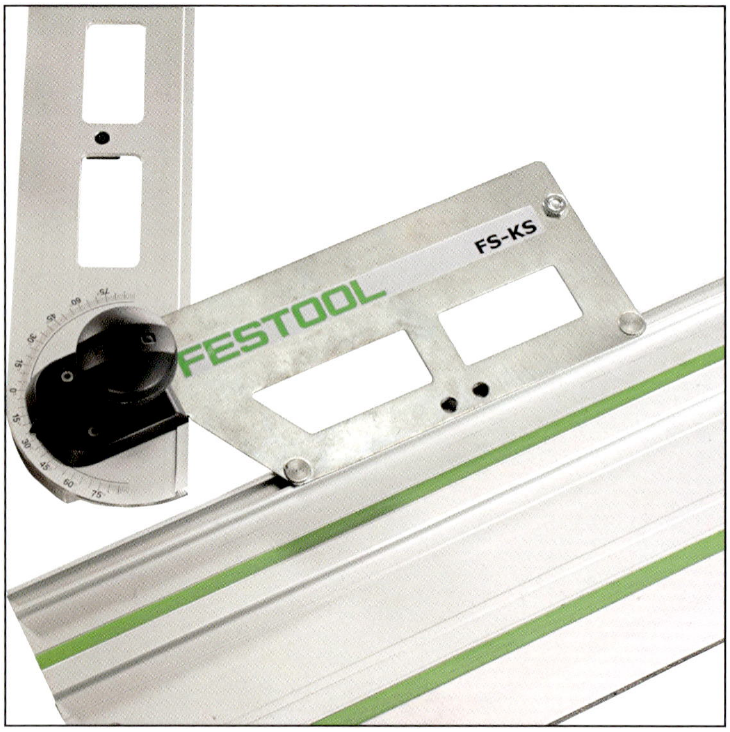

The FS-KS angle unit works with the guide rail to allow for accurate angle cutting.

These keys in the bottom of the angle unit, run in the channel on the surface of the guide rail.

Place the angle unit in the channel and check for snugness. You want the unit to slide easily in the channel but with no play.

Each key has an adjustment screw.

To check for true, set at 0° and hold a square in the angle.

Loosen the two screws on top of the indicator and adjust as needed.

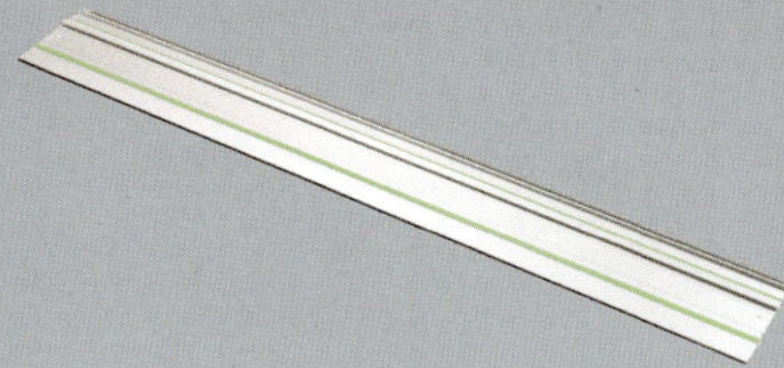

Guide Rail Accessories

The underside of the fence rides on the work piece.

To use, set the desired angle, place the angle unit in the guide rail, and position on the work piece.

Place the saw on the guide rail and make the cut.

LIMIT STOPS

A limit stop can be attached to the guide rail to act as a positive stop for the tool. It can attach to either extrusion.

One application is with the router for creating stopped flutes.

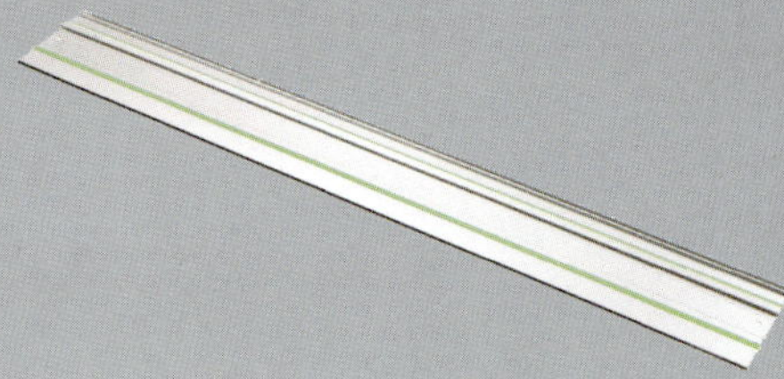

THE GECKO

The Gecko is a lever-activated double suction cup material holder. Designed to work on non-porous surfaces, it has a number of uses, including carrying materials, and can lift or carry loads up to 110 lbs. With the guide rail attachment it is used to secure the guide rail to non-porous flat surfaces.

The underside has two large suction cups.

The Gecko with guide rail attachment rides in the outside channel of the guide rail.

Tighten the Gecko to the rail.

USING THE GUIDE RAIL WITH OTHER FESTOOL TOOLS

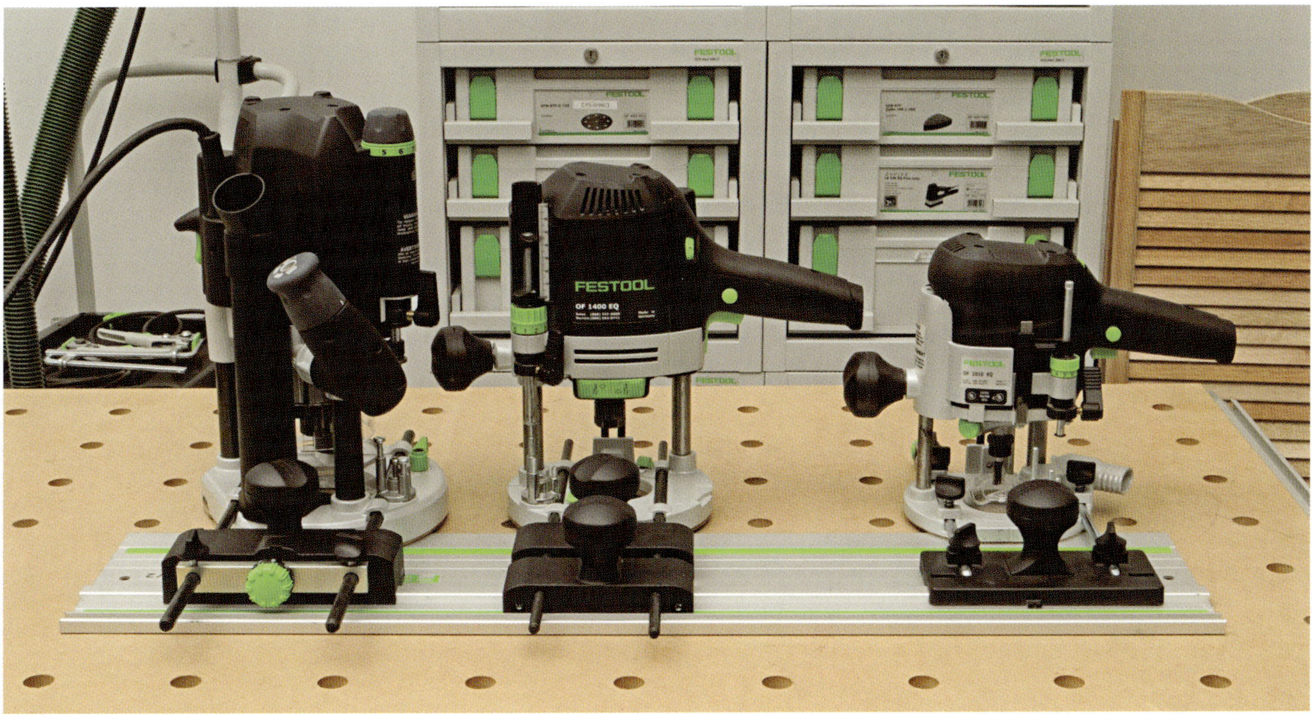

The guide rail works with all three Festool OF routers, left to right the OF 2200, The OF 1400 and the OF 1010.

Both Festool PS jigsaws use the same snap-in plate to connect to the rail system. Both tools are the same power but with different configurations. This is the barrel grip model PS 300.

The top handle model is the PSB 300.

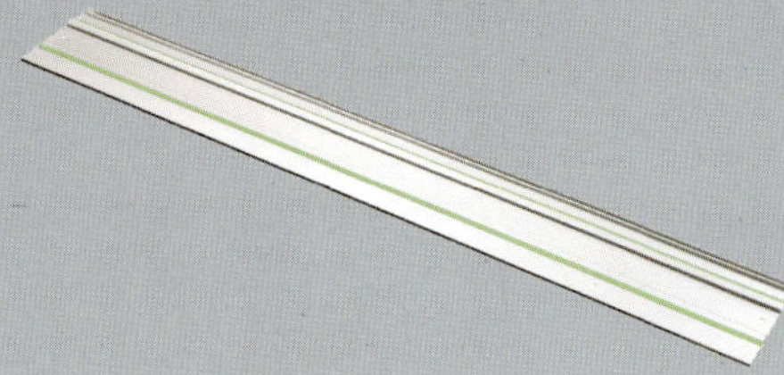

Applications of the Guide Rail System

One of the frustrations of most portable circular saws is the difficulty of getting an accurate, straight cut. It involves a complicated alignment procedure, using a typical straight edge, measuring from the blade to the edge of the foot plate, and praying that the edge is straight and the saw doesn't wander away from the line. With the Festool guide rail system everything is made simple. The key to this is the splinter guard is an accurate indicator of the cutting edge and can be aligned directly on the scribe mark. When used with the TS plunge saw, this is true for a beveled cut as well as a straight cut.

Angled Square Cut

Because of the accuracy of the splinterguard, a simple angled square cut can be made by laying the guide rail between two marks.

Suppose an angled cut is needed, going from 30mm at one edge...

...to 20mm at the other.

Applications of the Guide Rail System

Align the splinter guard on one mark...

...and the other...

Applications of the Guide Rail System

The result is a beautiful, clean cut right to the scribe line.

One application for the guide rail system is ripping heavy stock. This often involves special jigs, saws, or, for joining, special planers. With the Festool guide rail system the saw is taken to the wood in a relatively simple procedure. In this example a rip cut is made in an 8 foot piece of 8/4 stock. The piece is first clamped to the table using ratcheting clamps from under the table top. Even with the deep overhang I need, this clamping securely holds the work.

Applications of the Guide Rail System

A long guide rail is aligned on the scribe marks. In this case the goal is to clean up the edge.

To fix the guide rail to the wood a clamp is inserted into the bottom channel.

It then slides over to the work piece and is tightened, securely engaging the guide rail. Clamp from both ends.

Applications of the Guide Rail System

The TS 75 saw is used on this cut, though the TS 55 could be used with the proper ripping blade. The bigger saw, of course, handles it much more easily. Set the blade depth so it cuts through the stock.

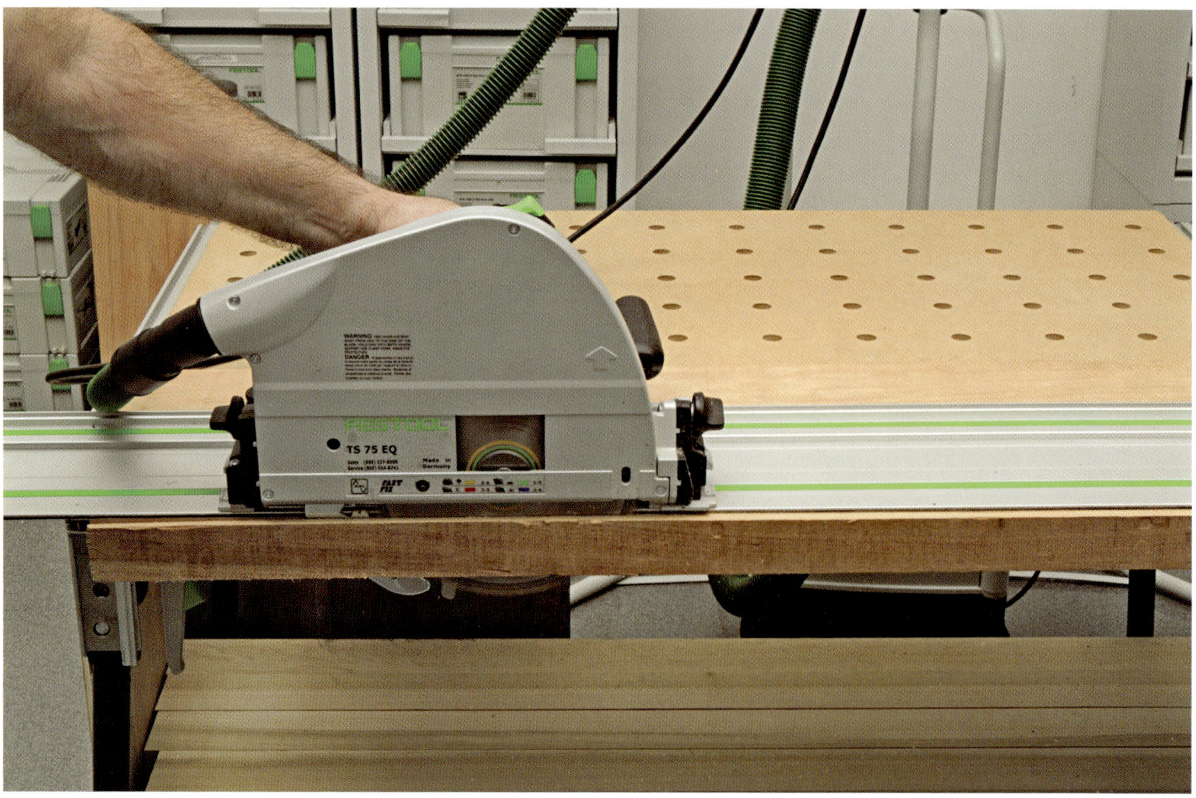

For ripping, plunge the blade before entering the material.

Applications of the Guide Rail System

The saw follows the guide all the way through the cut.

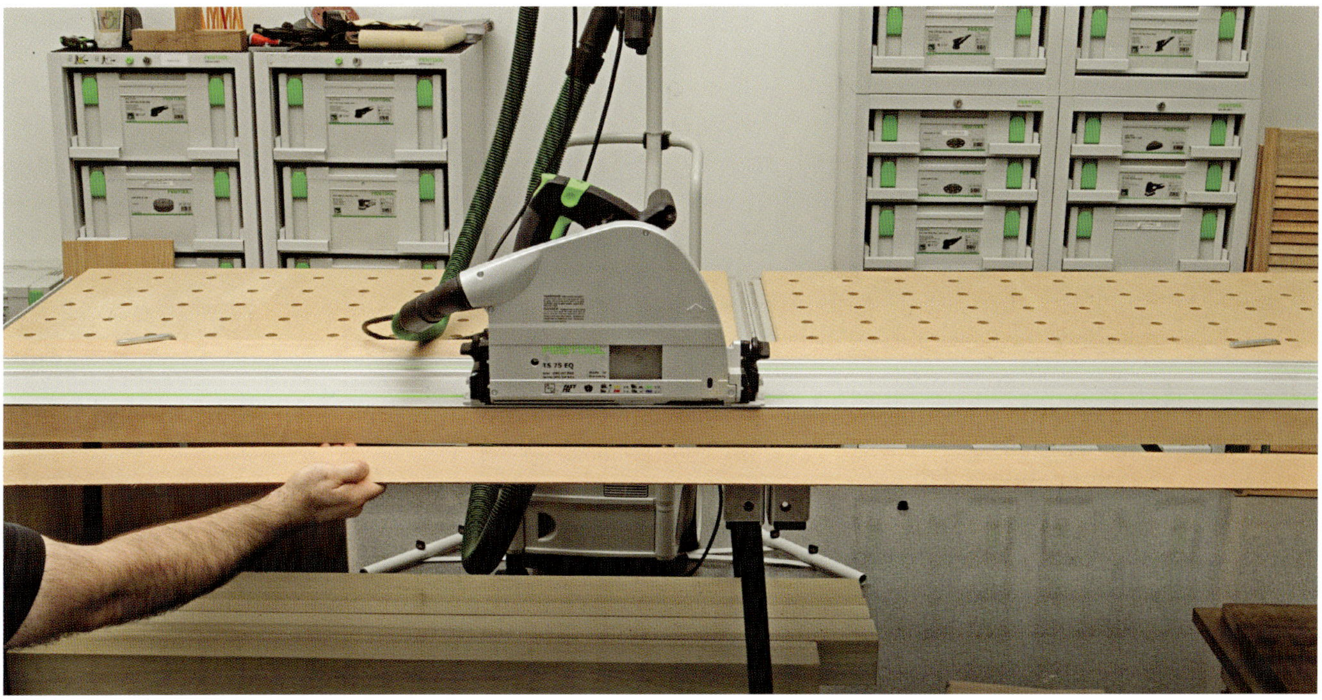

The result, even in this 8/4 cherry board is a finished cut throughout. There is no burning...

...and virtually no saw marks. In short it is a joinable edge, ready for gluing.

4. THE DUST EXTRACTION SYSTEM

Dust control is an integral part of the Festool system and design philosophy. It offers a full line of dust extractors designed to for the workshop or job site, to work with the Festool tools to optimize their operation and effectiveness at the same time as they provide a safe, clean work place.

The cleaning sets come in metal or plastic and the choice of anti-static or non anti-static hoses in three diameters to meet all your tool and dust extraction requirements. A wide assortment of connectors, reducers, and special sleeves allow you to tailor your system.

Festool offers a wide selection of filtration options for ultimate flexibility. In addition to the standard disposable filter bags, Festool offers a long life, re-usable bag. There are four main filters to choose from; standard, long-lasting, wet and HEPA (only available on CT 22 and CT 33). An optional dirt trap is available for wet applications.

Sys-Docks and optional hose hooks, holders and garages (for the CT22 and CT33) are designed for efficient storage of cleaning accessories and Festool Systainers. The flat tops offer added space for tools while you work.

With the optional boom arm, the hoses and cords are kept up and out of the way, opening up the work space.

The CT 22 and CT 33 can be equipped for connecting to pneumatic tools. Compressed air service units, auto-start conversion kits and IAS2 single and double adapters make the CT even more versatile.

Features

· **Tool-triggered activation** for automatic dust extraction.

· **Flat tops with locking latches** to attach Systainers, for ease of transport and storage.

· **Simple, pressure-fit hose connectors** for switching tools quickly.

· **Antistatic hoses are standard** to reduce dust build-up.

· **Ergonomically designed carrying handles and oversized wheels** for ease of movement around the job site.

· **Built-in cord storage** reduces tripping hazards.

· **HEPA filters** eliminate 99.97% of particulates from the air (Only available on CT 22 and CT 33, where they are standard equipment)

· **The motor is completely separate from the dust bin,** significantly prolonging the life of the motor.

· **Adjustable cubic feet per minute (CFM)** setting for better results.

Specifications

Model name	CT MINI	CT MIDI	CT 22 E	CT 33 E
Item	**583 360**	**583 376**	**583 366**	**583 368**
Standard package includes				
Filters	1 micron	1 micron	HEPA	HEPA
Filter bag	2	2	1	1
Antistatic suction hose	1-1/16" x 11.5'	1-1/16" x 11.5'	1-1/16" x 11.5'	1-1/16" x 11.5'
	(27mm x 3.5m)	(27mm x 3.5m)	(27mm x 3.5m)	(27mm x 3.5m)
Container/filter bag capacity	2.6 gal/2 gal	3.96 gal/3.3 gal	5.8 gal/5.3 gal	8.7 gal/7.9 gal
	10 liters/7.5 liters	15 liters/12.5 liters	22 liters/20 liters	33 liters/30 liters
Specifications				
Depth x width x height	17.5" x 13.5" x 16.5"	17.5" x 13.5" x 18.5"	24" x 15" x 17"	24" x 15" x 22"
	(440 x 340 x 420 mm)	(440 x 340 x 470mm)	(625 x 380 x 440 mm)	(625 x 380 x 550 mm)
Weight	21 lbs. (9.5 kg)	21.4 lbs. (9.7 kg)	26.5 lbs. (12 kg)	32 lbs. (14.5 kg)
Noise (sound level)	62 dB (low setting)	62 dB (low setting)	62 dB (low setting)	62 dB (low setting)
	72 dB (maximum setting)	72 dB (max. setting)	72 dB (maximum setting)	72 dB (maximum setting)
Maximum suction capacity	99 CFM	99 CFM	134 CFM	134 CFM
	(2807 l/min)	(2807 l/min)	(3800 l/min)	(3800 l/min)
Maximum vacuum	80" static water lift	80" static water lift	90" static water lift	90" static water lift
	(200000 Pa)	(20000 Pa)	(23000 Pa)	(23000 Pa)
Power consumption	400-1200 watts /	400-1200 watts /	350-1200 watts /	350-1200 watts /
	3.3-10 amps 120 v AC	3.3-10 amps 120 v AC	2.9-10 amps 120 v AC	2.9-10 amps 120 v AC

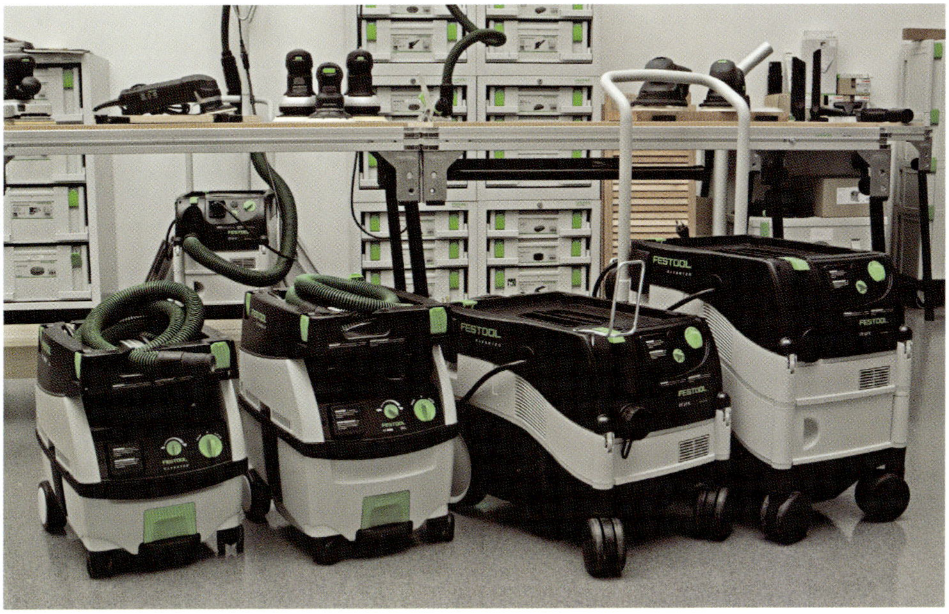

Left to right: CT Mini (99 cfm, 2 gal. cap.), CT Midi (99 cfm, 3.3 gal. cap.), CT-22 (134 cfm, 5.3 gal cap.), and CT-33 (134 cfm, 7.9 gal. cap.). The smaller two units are great for job sites, being easily portable and small enough to get in tight spots. The CT 33 is mainly a workshop unit, with large capacity and power. The CT 22 is an all purpose unit, with power and size that are good for the shop and job site.

Description/Application	Item	Compatability			
		CT MINI	CT MIDI	CT 22 E	CT 33 E
Hoses—Antistatic					
Antistatic versions are only effective when used with antistatic design (Festool's CT series) dust extractors.					
With rotating connector and reducing sleeve.					
1 1/16" x 11.5' (27 mm x 3.5 m) AS	**452 878**	●	●	●	●
1 1/16" x 16.5' (27 mm x 5 m) AS	**452 880**	●	●	●	●
1 7/16" x 11.5' (36 mm x 3.5 m) AS	**452 882**			●	●
1 7/16" x 16.5' (36 mm x 5 m) AS	**452 884**			●	●
1 7/16" x 23' (36 mm x 7 m) AS	**452 886**			●	●
1 15/16" x 13' (50 mm x 4 m) AS	**452 888**			●	●
1 15/16" x 13' (50 mm x 4 m) AS	**452 890**			●	●
Hoses—Non antistatic					
For use with cleaning sets or non-antistatic dust extractors.					
With rotating connector and reducing sleeve.					
1 1/16" x 11.5' (27 mm x 3.5 m)	**452 877**	●	●	●	●
1 1/16" x 16.5' (27 mm x 5 m)	**452 879**	●	●	●	●
1 15/16" x 11.5' (36 mm x 3.5 m)	**452 881**			●	●
1 15/16" x 16.5' (36 mm x 5 m)	**452 883**			●	●
1 15/16" x 23' (36 mm x 7 m)	**452 885**			●	●
1 15/16" x 8.25' (50 mm x 2.5 m)	**452 887**			●	●
1 15/16" x 13' (50 mm x 4 m)	**452 889**			●	●
Hose Sleeves—Rotating connector					
Hose connector to fit inlet of dust extractor, external dia. 2 9/32" (58 mm).					
Antistatic version for D 50 suction hose	**452 896**			●	●
Antistatic version for D 36 suction hose	**452 894**			●	●
Antistatic version for D 27 suction hose	**452 892**			●	●
Non-antistatic version for D 50 suction hose	**452 895**			●	●
Non-antistatic version for D 36 suction hose	**452 893**			●	●
Non-antistatic version for D 27 suction hose	**452 891**	●	●	●	●
Hose sleeves--Reducing sleeves					
Hose connector for tools or cleaning sets.					
With rotating adapter for D 27 suction hose	**487 071**	●	●	●	●
Replacement. Tool end of extraction hose for our standard 27mm hose.					
Antistatic version with rotating adapter for D 36 suction hose	**487 721**			●	●
Replacement. Tool end of extraction hose for mid-sized 36mm hose.					

Accessories

Description/Application	Item	CT MINI	CT MIDI	CT 22 E	CT 33 E
		Compatability			
Non-antistatic hose reducer *For reducing from 50mm to 36mm hose or when connecting 36mm to 27mm hose.*	**452 897**			●	●
Antistatic angle adapter *Elbow hose end for connection in CT MIDI or CT MINI*	**456 806**	●	●		
Antistatic connector sleeve *To connect suction hoses D 50/D 36/D 27.*	**493 047**			●	●
Antistatic Y-piece with blanking plug *To connect two suction hoses to the dust extractor.*	**452 898**			●	●
Blanking plug *For retrofitting on CT 23/33, to seal off inlet during transportation.*	**452 899**			●	●
Tubes—Curved					
Stainless steel *To connect tubes to extraction hose for comfortable grip.*	**452 900**	●	●	●	●
Polypropylene (PP) with air regulating slide to adjust suction power *To connect tubes to extraction hose for comfortable grip.*	**452 901**	●	●	●	●
Tubes—Extension					
Stainless steel 3 piece, 37 1/2" (950 mm) D 36	**452 902**	●	●	●	●
Non-antistatic polypropylene (PP) 3 piece, 37 1/2" (L= 970 mm) D 36	**452 903**	●	●	●	●
One-piece anodized aluminum 43 1/2" (L= 1100 mm) D 50	**447 599**			●	●
Non-antistatic plastic 2-piece, 38" (L= 970 mm) D 50	**440 412**			●	●
Floor nozzles/brushes					
Industrial floor nozzle *High-quality industrial design; aluminum.* *Width 14 5/8" (370 mm); 2 casters, 2 brush inserts.*	**452 908**	●	●	●	●
Large Industrial floor nozzle *High-quality industrial design; aluminum.*	**452 910**	●	●	●	●

Accessories

Description/Application	Item	CT MINI	CT MIDI	CT 22 E	CT 33 E
Width 17 11/16" (450 mm); 2 casters, 2 brush inserts.					
Standard floor nozzle	**452 907**	●	●	●	●
Plastic. Width 11 13/16" (300 mm) 2 caster, 2 brush inserts					
Workshop floor nozzle	**452 909**	●	●	●	●
Plastic. Width 14 5/8" (370 mm) 2 caster, 2 brush inserts					
Interchangeable floor nozzle	**452 906**	●	●	●	●
4-piece, plastic. With 3 replacement inserts (brushes, rubber lips, carpet)					
Multi-purpose floor nozzle	**452 911**	●	●	●	●
With retractable brush inserts.					
Turbo suction brush	**450 644**			●	●
For carpet and hard surface with 4 casters. Width 10 5/8" (270 mm). Suction-activated rolling brushes, requires 36 mm hose.					
Special nozzles/brushes					
Suction brush	**440 404**	●	●	●	●
Plastic, dia. 2 13/16" (70 mm)					
Upholstery brush	**440 406**	●	●	●	●
Plastic, with brush insert width 4 11/16" (120 mm)					
Universal brush nozzle	**440 403**	●	●	●	●
Plastic, brush ring 4 11/16" x 1 5/8" (120 x 40 mm)					
Large crevice nozzle	**452 912**			●	●
Plastic, length 14 4/8" (370 mm)					
Crevice nozzle	**452 904**	●	●	●	●
Plastic, length 14 5/8" (370 mm)					
Bevel ended nozzle	**411 810**	●	●	●	●
Flexible rubber, length 7 1/2" (190 mm), outside diameter 1 5/8" (40 mm)					
Other accessories					
Spark trap	**484 733**			●	●
For use when operations may generate sparks, i.e. sanding or sawing metal.					
Handle	**452 921**			●	●
For CT 22 & CT 33 only. Improves mobility. Required for mounting Boom Arm 492 753.					
Suction hose holder	**487 072**			●	●
Raises hose to table height and prevents trip hazard.					
Hose hook	**452 998**			●	●
Hose minder for storage or to prevent hose from laying on the ground.					

(Compatability: CT MINI, CT MIDI, CT 22 E, CT 33 E)

Accessories

Description/Application	Item	CT MINI	CT MIDI	CT 22 E	CT 33 E
		Compatability			
Dirt trap *Collection bucket facilitates disposing of liquids.*	**452 925**			●	
Dirt trap *Collection bucket facilitates disposing of liquids.*	**452 926**				●
CT adapter plug *Replacement. Adapts the 20 amp CT plug to fit in standard 15 amp outlets.*	**493 232**	●	●	●	●
Boom arm *Keeps the cord and hose up and out of your way; requires handle (SB-CT 452 921) to be mounted onto CT 22/33 dust extractor.*	**492 753**			●	●
CT hose garage *Stores and protects hose and cord for transport.*	**494 388**			●	●
Boom Arm Tool Holder *Convenient storage and cradle of two electric sanders. For use with Boom Arm only. Includes wire rack and clamping knob.*	**493 558**			●	●
Filter Elements					
Main filter element *Standard 1 micron filter replacement-paper. 2 pieces.*	**452 923**			●	●
HEPA filter element *The highest standard for filtration. Filters particles to .3 microns.*	**493 334**			●	●
Wet filter element *Prevents foam buildup associated with wet extraction. 1 piece.*	**454 805**	●	●		
Wet filter element *Prevents foam buildup associated with wet extraction. 2 pieces.*	**452 924**			●	●
Longlife filter element *Cloth filter with special "M" fold prevents build-up. 2 pieces.*	**454 869**			●	●
Filter bags					
Filter bag *Paper filter bag captures dust. 5 pieces.*	**456 772**	●			
Filter bag *Paper filter bag captures dust. 5 pieces.*	**494 105**		●		
Filter bag *Paper filter bag captures dust. Includes dust cap for clean disposal. 5 pieces.*	**452 970**			●	

Accessories

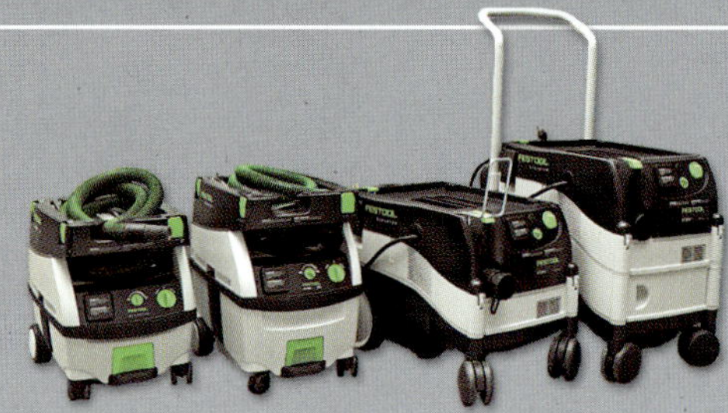

Description/Application	Item	Compatability			
		CT MINI	CT MIDI	CT 22 E	CT 33 E
Filter bag *Paper filter bag captures dust. Includes dust cap for clean disposal. 5 pieces.*	**452 971**				●
Bulk filter bags *20 pieces.*	**494 631**			●	
Bulk filter bags *20 pieces.*	**494 632**				●
Longlife filter bag (re-usable bag) *Economical solution for capturing and eliminating large chips produced by routing or planing. (at least 500 fillings) Re-sealable.*	**456 737**			●	
Longlife filter bag (re-usable bag) *Economical solution for capturing and eliminating large chips produced by routing or planing. (at least 500 fillings) Re-sealable.*	**456 738**				●

The CT MINI & MIDI

The **CT MINI** and **CT MIDI** are Festool's most portable mobile dust extractors. Compact and lightweight, they have enough capacity for common job site tasks. The CT MIDI, somewhat larger in size, offers increased capacity for heavier on-site use. Both units have large wheels and double casters, that can easily negotiate obstacles in the shop or on the job site. Generating 99 CFM of suction, and only 62 dB (on low setting), these CTs can quickly and quietly remove large amounts of debris. One disposable filter comes standard on the CT MINI and CT MIDI, with the option of additional dry and wet filters to meet your every need. The integrated hose garage keeps hoses and cords well organized and makes Systainer transport simple and efficient.

CT Mini & CT Midi Features:

• **Integrated Sys-Dock and hose garage.** Cords and hoses conveniently store in the onboard hose garage, while the flat top and locking latches allow you to attach one or more Systainers.

• **Filter bag.** Disposable, two-ply filter bags are particularly tear-resistant; an integrated dust cap prevents dust from spilling during disposal.

• **Tool-less main filter and automatic water level detector.** The main filter can be changed without using tools. The internal sensors monitor the fluid level, for wet applications, and switch the mobile dust extractor off automatically when the container is full.

• **Automatic tool start/Adjustable suction.** Smooth start, infinitely adjustable electronic air flow control and tool-triggered or manual on-off switch ensures convenient and efficient dust control.

• **Convenient carry handle.** The ergonomic handle makes maneuvering these small and light dust extractors a breeze.

• **Foot brake.** Quick on/off foot actuated brake prevents roll-off during usage or during transport.

• **Anti-static hose.** A 1" x 11.5' (27mm x 3.5m) anti-static hose comes standard.

• **Large wheels.** Large wheels, low center of gravity, and ergonomically balanced design allows you to move the CT effortlessly around the job site.

• **Mobility.** The manageable size makes it easy to transport or carry up stairs.

The CT MINI & MIDI

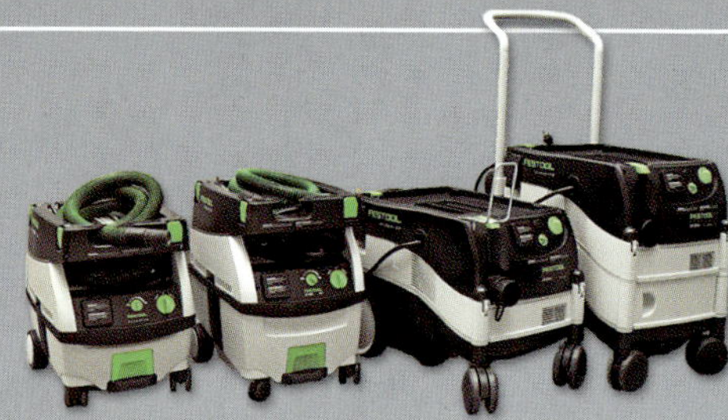

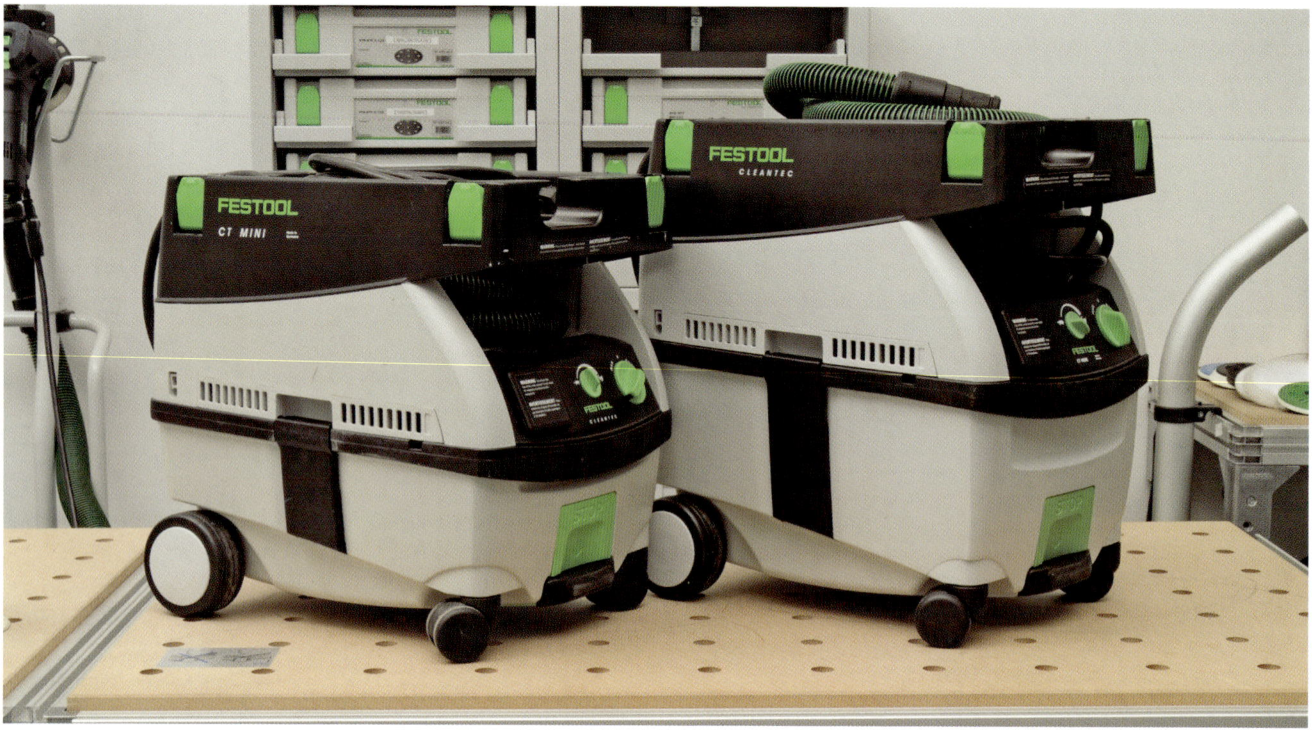

The CT MINI (left) and CT MIDI (right) are nearly identical machines, except for the larger capacity of the MIDI.

Foot brake. When pushed all the way down, it raises the wheels off the ground and keeps the unit in place. This is useful during transport of the unit and on the job site.

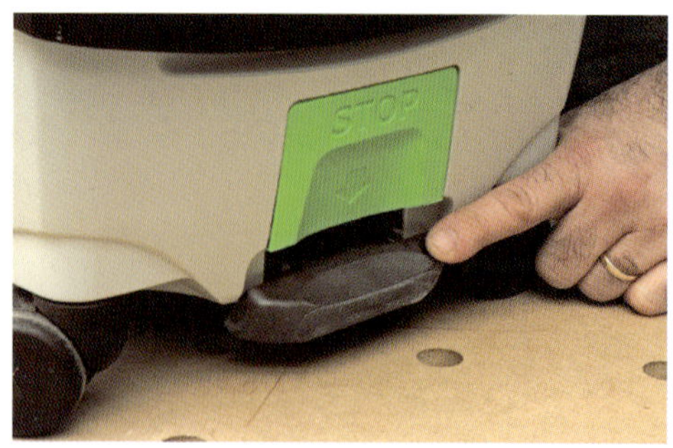

The 27 mm anti-static hose does not allow dust to cling during use. When the vac is turned off, there is no dust in the hose to fall back into the workspace.

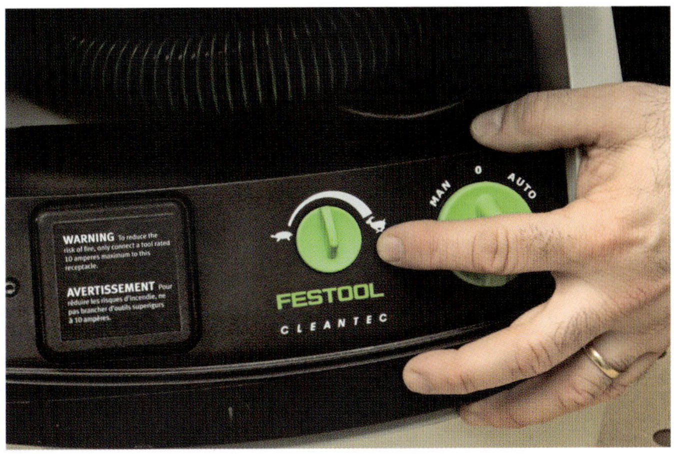

Variable suction control. This is especially important during sanding operations, where a powerful suction could pull the pad too snugly against the work. This would create excessive friction and heat, making swirl marks on the material. By lowering the suction to the correct level, the sanding proceeds much more smoothly and cleanly.

ON/OFF switch. "0" equals off.

"MAN" is manual, turning the vac on.

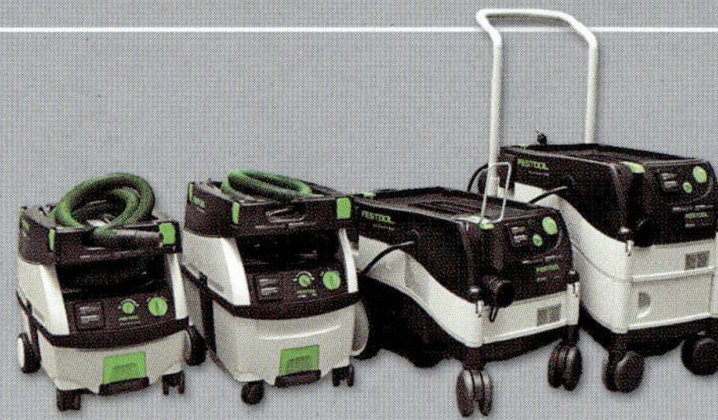

"AUTO" is automatic, meaning that the vac is turned on when the tool is on.

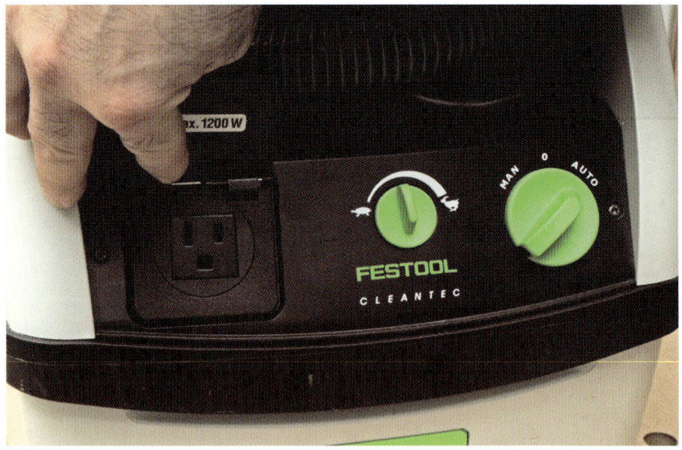

The AUTO function works when the tool is plugged into this outlet on the vac.

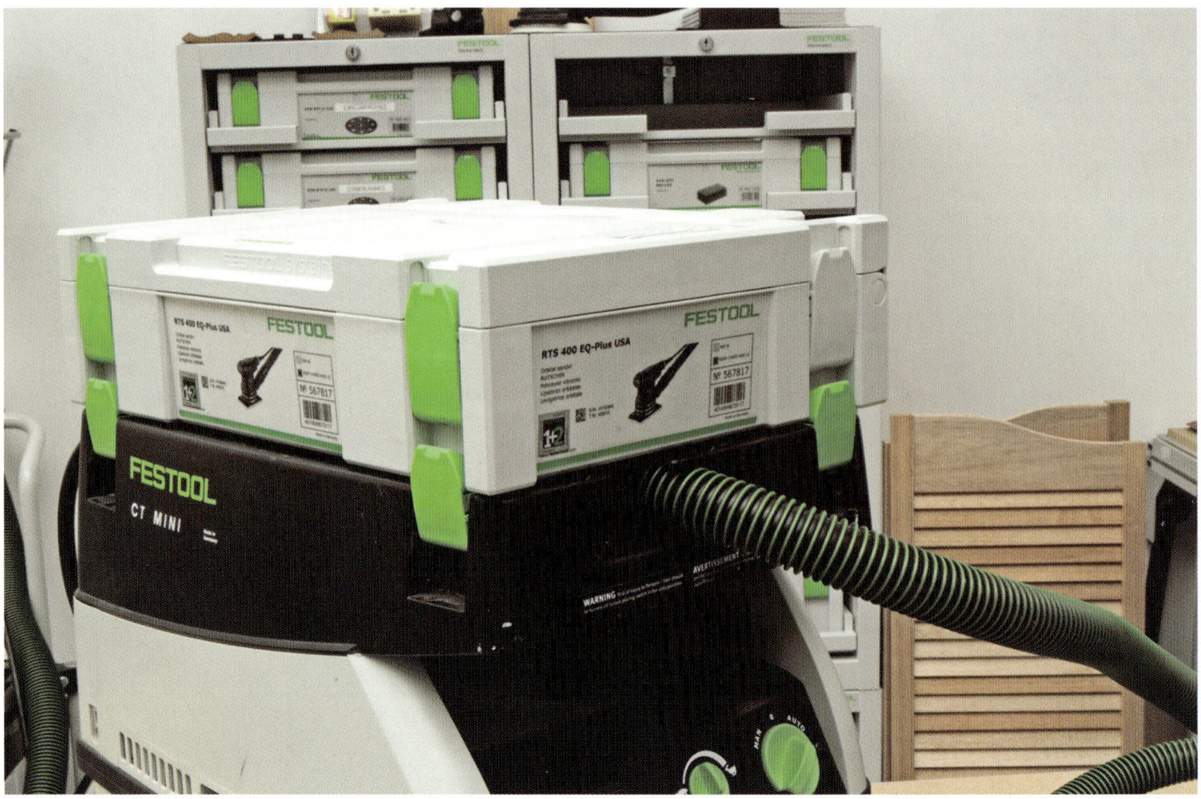

The SYS-DOCK allows various SYSTAINERS to be securely attached to the top of the vac for easy transport…the vac actual serves as a dolly.

The CT MINI & MIDI

There are four latches for the SYS-DOCK, one at the front and back, and two on one side. With the SYSTAINER in place, open the latch, pull it up and lock it in place.

To gain access to the bag and filters, undo the latches on both sides of the base.

Lift the top by the handle.

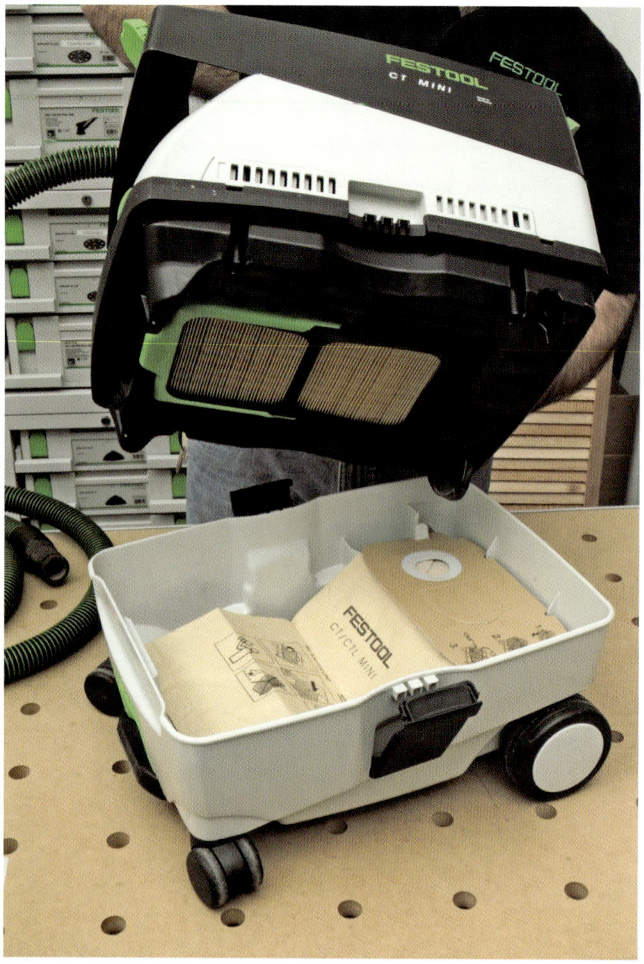

This reveals the filter bag, in the bottom compartment, and the filters under the lid.

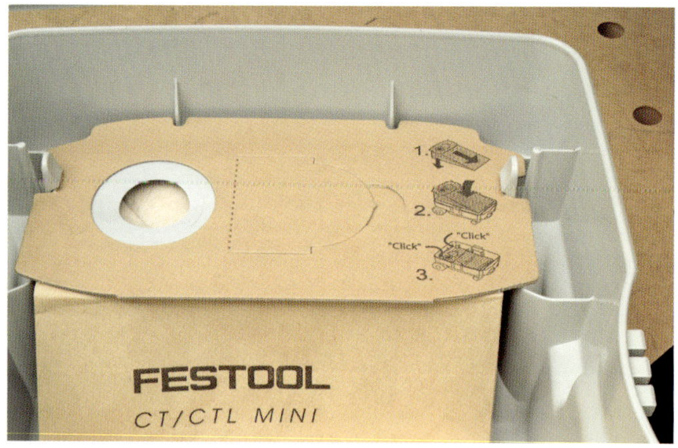

The bag in place. The dust comes from the job directly into the bag. Particles 5 microns or greater stay there. The smaller bits move through the bag into the filter, which captures particles 1 micron in size or larger.

Clips on either side hold the bag in place.

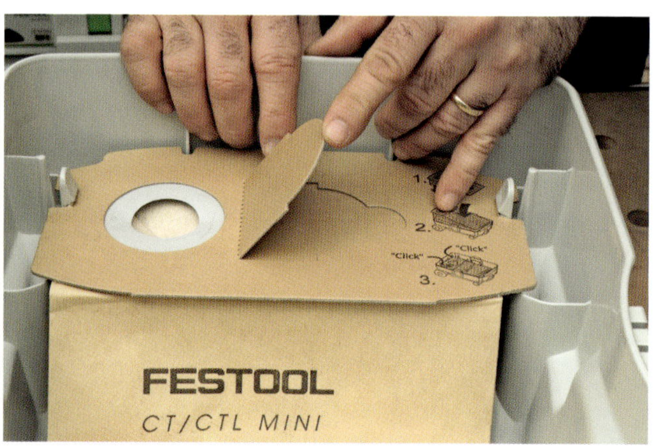

When the bag is full this tab is lifted…

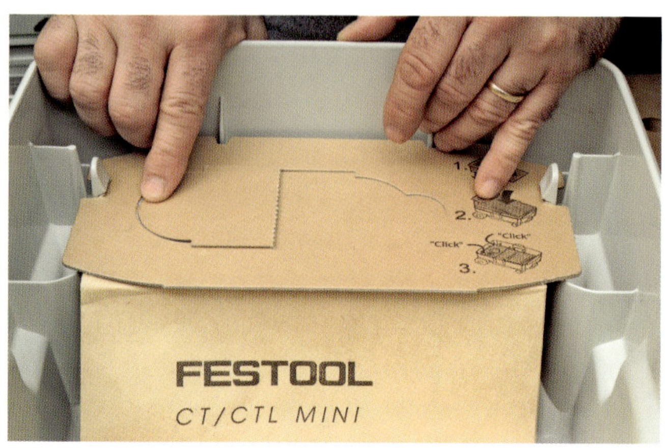

…and pressed in place to seal the bag, keeping the dust inside.

The 1 micron filter.

For cleaning, pull the handle out...

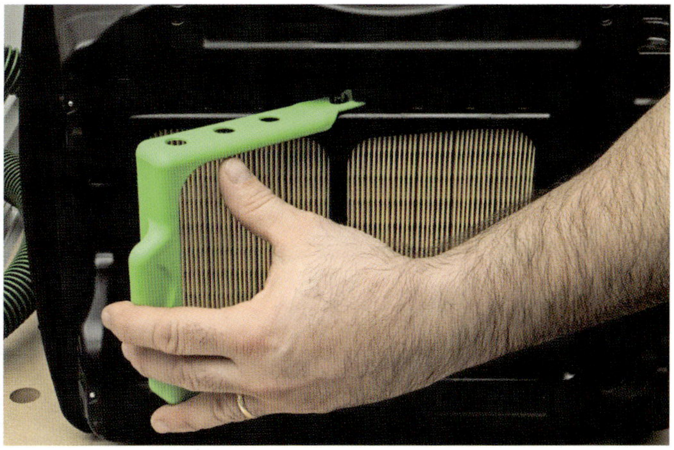

...and remove the filter.

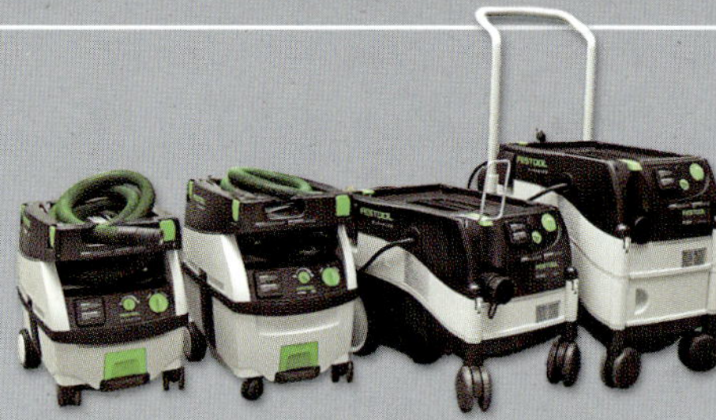

The CT MINI & MIDI

To clean, remove the filter from the frame…

…and tap against the side of a trash can. Used properly the filters will last a very long time. Replace when there is a noticeable loss of air flow, even after replacing the bag and cleaning the air filter.

Extra filter bags can be stored in this space at the back.

The CT MINI & MIDI

The hose and cord are neatly stored in the top.

The vacs come with a 20 amp plug to go into 20 amp outlets, now required in most commercial applications.

Since most residences and older construction do not have these outlets, there is a 15 amp plug adapter available. These are easily lost, so it is best to have a few in your toolbox.

The CT MIDI bag has a 3.3 gallon dry capacity, about 1.3 gallons more than the CT MINI.

For wet vac use, the filter can be replaced with a foam breaking wet filter element.

Remove the bag.

Converting the CT MINI & MIDI for Wet Vacuuming

Remove the 1 micron filter...

...and replace with the foam breaking filter.

There are two water level sensors (level stop sensors) that turn off the motor when the water gets to capacity.

CT 22 E & CT 33 E

The **CT 22 E** and **CT 33 E** are Festool's most powerful mobile dust extractors. With a sturdy chassis, extra large wheels, and double casters, these units can easily negotiate obstacles in the shop or on the job site. With 134 CFM and only 62 db (on the low setting), these CTs can quickly and quietly remove large amounts of debris, with minimal noise distractions. HEPA filters come standard on the CT 22 E and CT 33 E and offer the ultimate in clean air and filtration. Integrated Sys-Docks or the optional hose garage allow you to attach one or more Systainers for added convenience. An optional boom arm makes a perfect companion in a shop environment and keeps hoses and cords organized and away from work surfaces.

CT 22 E & CT 33 E Features

• **Bypass blower port.** The bypass blower port allows you to evacuate unpleasant odors and fumes from the job site, by attaching an additional hose.

• **HEPA filters.** HEPA filters are standard and remove 99.97% of particulate to .3 micron.

• **Automatic tool start and adjustable suction.** Smooth start, infinitely adjustable electronic air flow control and tool-triggered or manual on-off switch ensures convenient and efficient dust control.

• **Convenient carry handle.** Ergonomic handle is positioned 7-9 inches in from the body, which is the most natural position for lifting.

• **Filter scrubber.** The convenient scrubber handle removes the accumulation of dust and debris on the internal filter, without the need to open the machine and improves air circulation across the motor.

• **Sys-Dock and anti-static hose.** The flat top with locking latches allows you to transport tools and accessories securely and conveniently when used with a Festool Systainer. As with all CT dust extractors, the 1-1/16" x 11.5' (27mm x 3.5m) anti-static hose comes standard.

• **Automatic water level sensor and optional dirt trap.** With wet applications, the internal sensors monitor the fluid level and switches the mobile dust extractor off when the container is full (dirt trap available as an accessory).

• **Cord storage.** Everything needs its own storage space. The cord can be wound on the dedicated cord wrap to prevent it from getting in the way during transportation.

• **Tool-less filter changes.** If there is ever a need to change filters, or if you want to switch to a sponge filter for wet applications, changing the filter is very easy. The flat filter can be changed without using tools.

CT 22 E & CT 33 E

The CT 33 E is nearly identical to the CT 22 E with the exception of capacity.
The CT 33 E holds 2.6 gallons more than the CT 22 E.

CT 22 E & CT 33 E

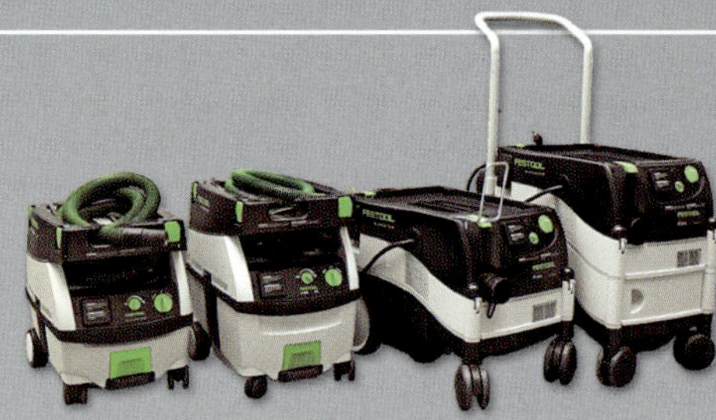

The CT 33 E has a small storage compartment in front, useful for spare bags.

The controls are the same as the smaller units…

…with the tool outlet behind the door.

The handle is ergonomically designed to be about 7-9 inches from the body, a comfortable carrying position.

The Sys-Dock system is somewhat different. Tabs at the lower edge of the Systainer...

...fit into sockets on the inner edge of the top of the vac.

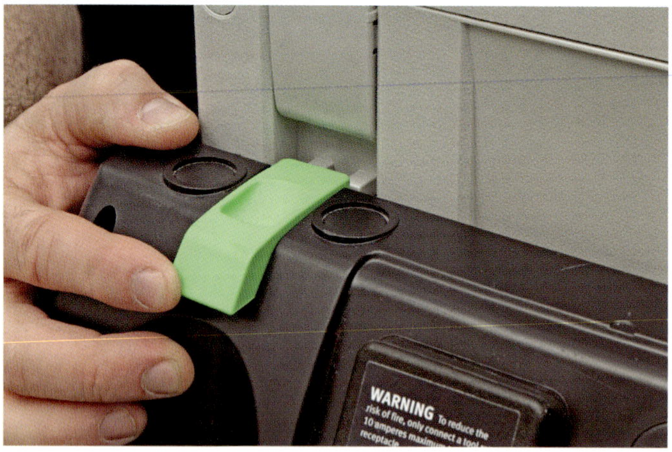

Lower the Systainer into place and latch it at the opposite side.

The latch simply slides into place.

Large, non-marring wheels on this unit allow for easy movement.

CT 22 E & CT 33 E

Wheel locks are located inside the rear wheels. Simply push them down to engage the lock.

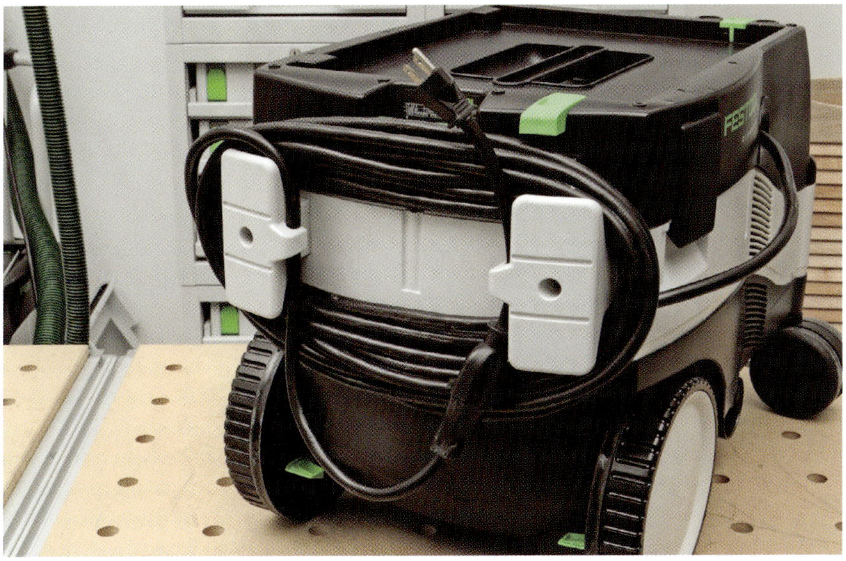

The cord is stored on the rear.

On the side are the exhaust ports. This grille is the normal exhaust outlet.

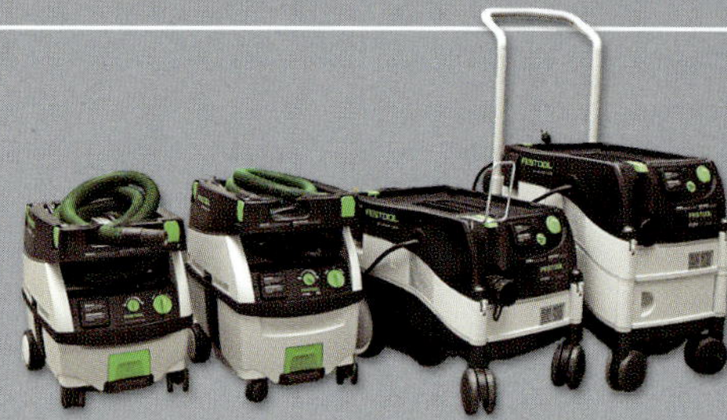

Beneath it is the round exhaust port that allows venting to the outside, removing any odors or vapors that are not captured by the HEPA filter. There is a flap inside that raises when the hose is inserted, blocking the normal venting system.

The exhaust hose can be run to a place outside the work area.

The green handle at the back of the unit operates the quick filter scrubber. Pull it in and out a couple of times while the unit is off and it knocks the dust into the reservoir. This will prevent the filters from caking. A thorough cleaning is done by removing the filters and tapping them.

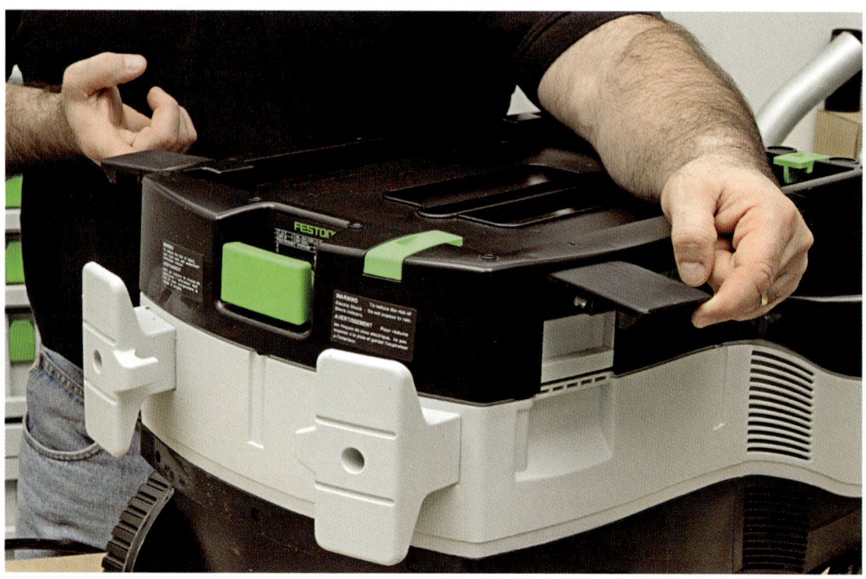

To access the bag and filters, lift the release latches on both sides.

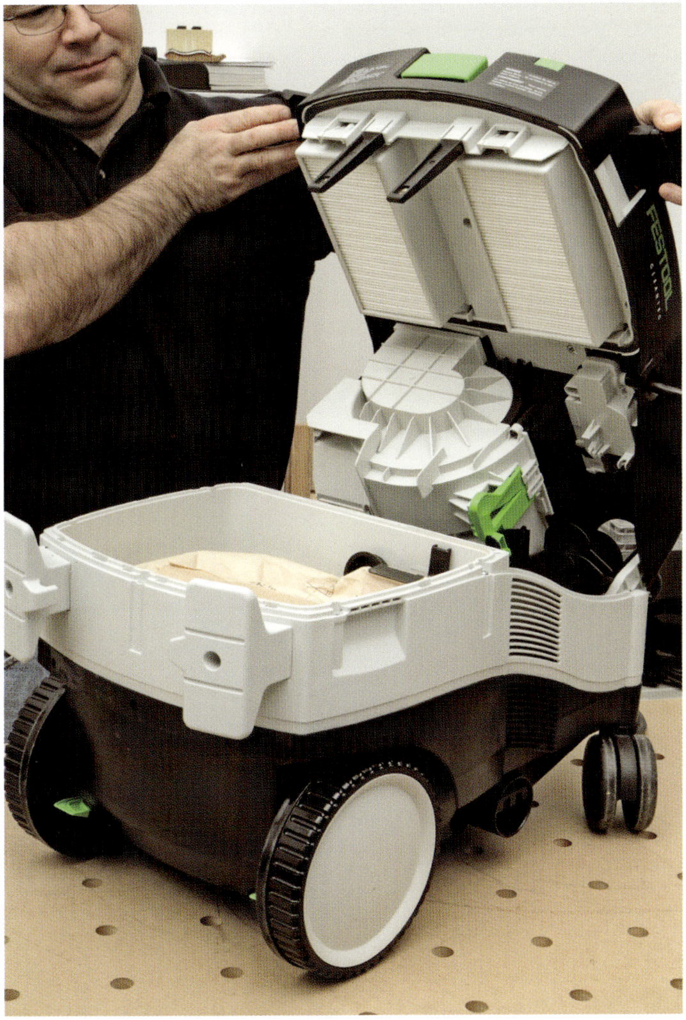

The lids lifts from the back.

A kickstand automatically engages to keep the lid up.

To release the kickstand, simple press in the tab.

CT 22 E & CT 33 E

The 5 micron bag fits in the well.

To remove it, pull back on the tab at the opening.

Insert the cap to seal the bag.

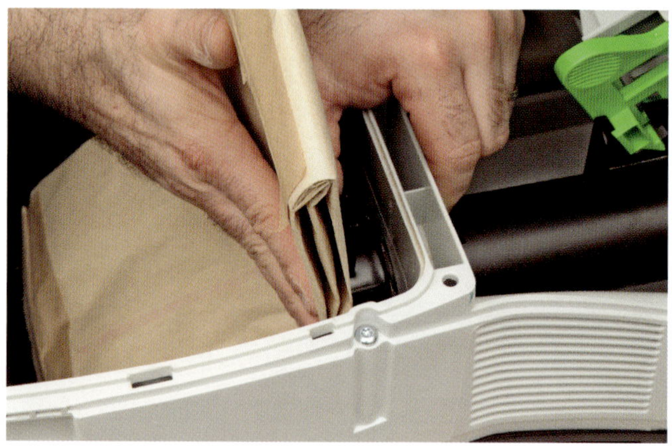

Insert a new bag, using your hand to keep it straight.

When it is snuggly in place, bend the tab down. This promotes uniform filling and prevents tearing when removing the bag.

This is the grounding contact for the hose. The grounding strap allows the unit to have continuity for the anti-static hose design and ground the intake port to the electrical ground.

The latch for the HEPA filter.

Depress the latch and pull down the filter frame.

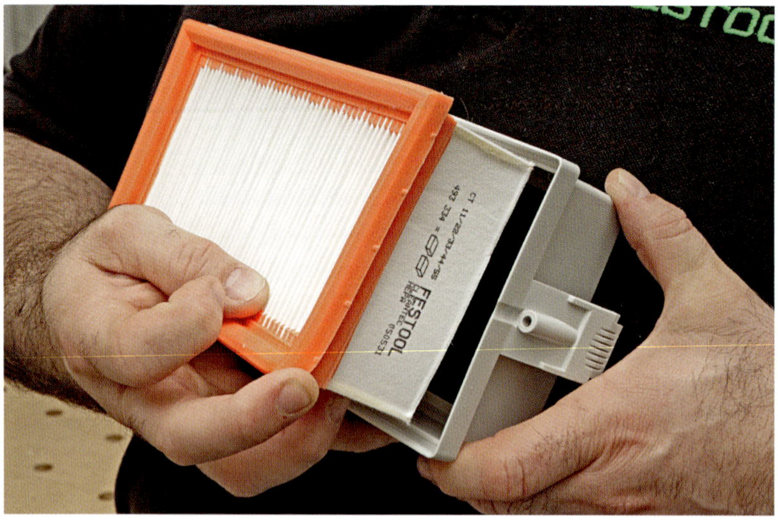

When the frame is free from the unit, the filter simply pops out.

Tabs at the front of the filter frame fit into slots in the unit.

Insert them and pivot the frame into position. The black apparatus above the filter tray is the quick filter scrubber.

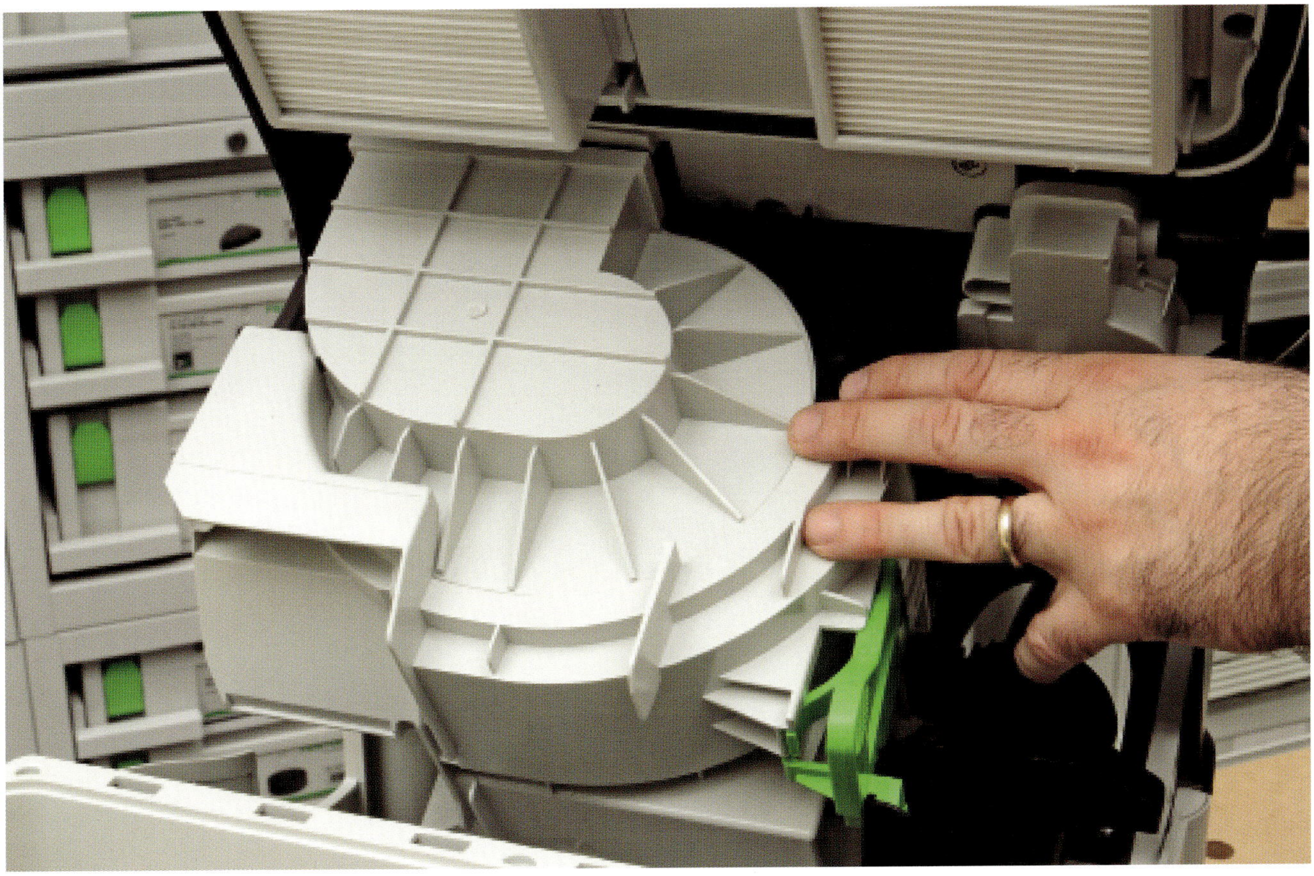

The motor is completely encased, keeping it free of dust and ensuring a long life.

In wet vac mode, the level stop sensors turn the motor off when the reservoir is filled.

Wet Vacuuming with the CT 22 E & CT 33 E

To convert to wet vac mode, remove the bag.

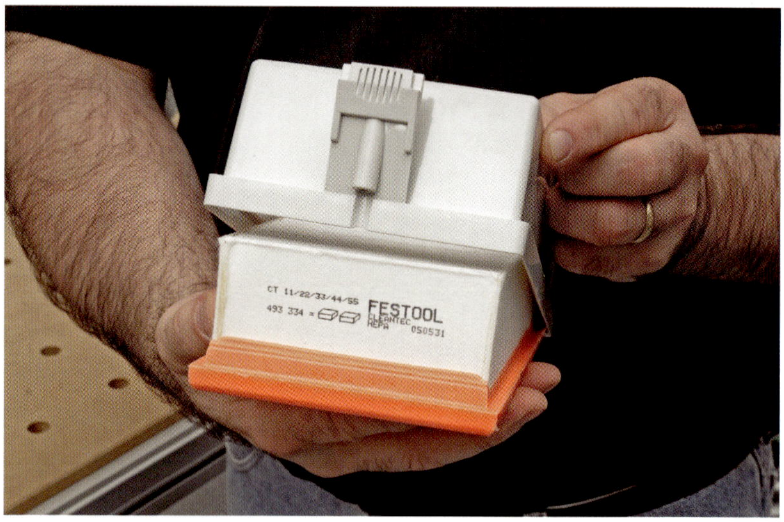

Remove paper filter…

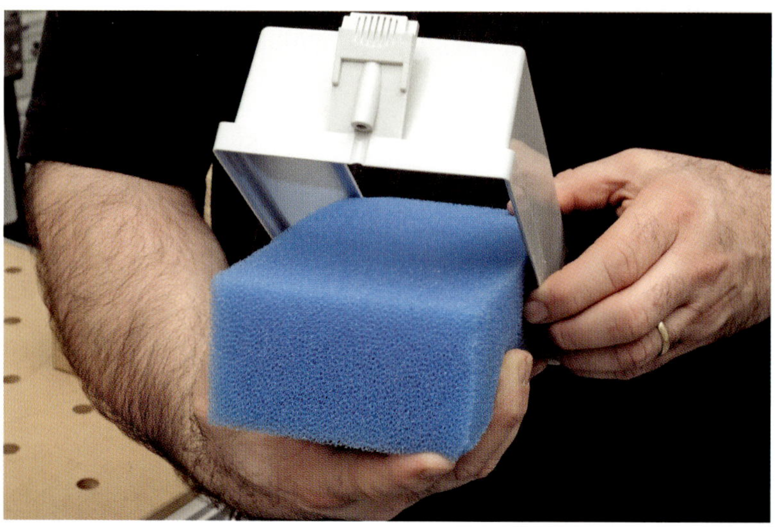

…and replace with foam-breaking wet filter element.

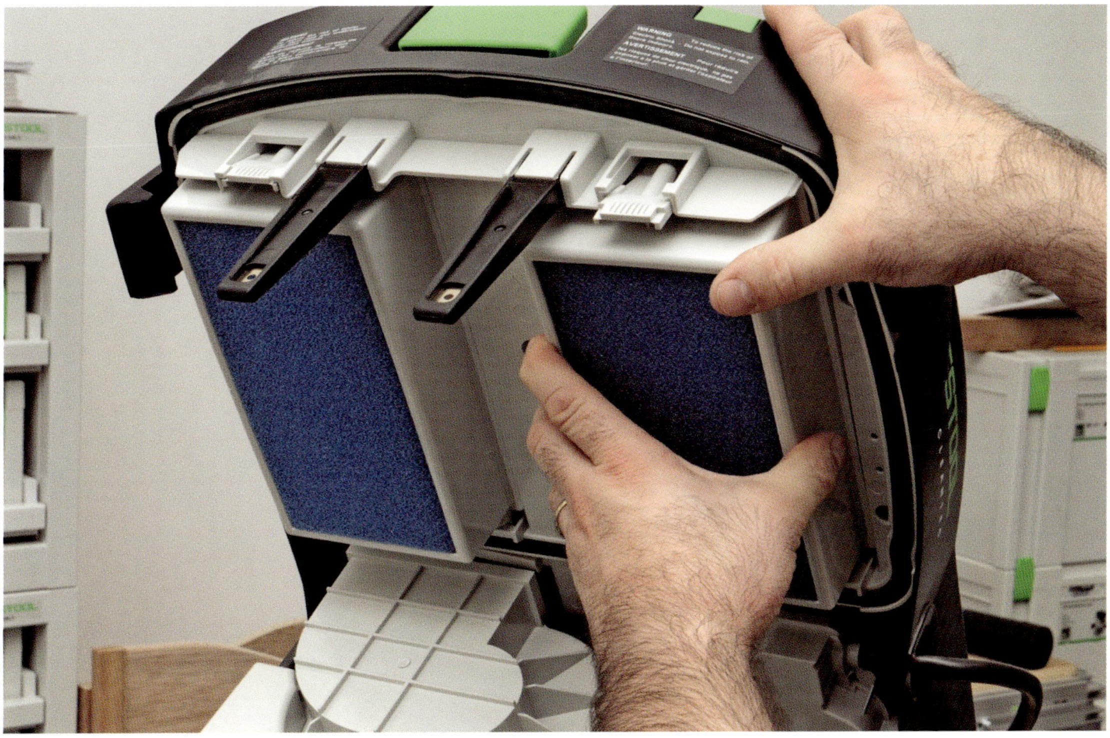

Reinsert the filter frames.

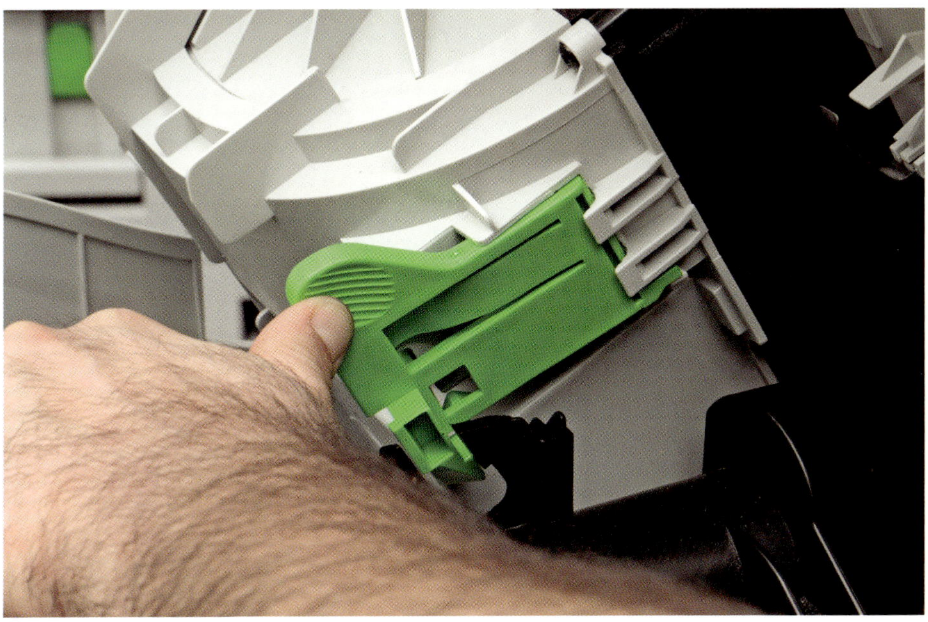

When the reservoir is full you can remove the top to make emptying it easier. Release the green kickstand.

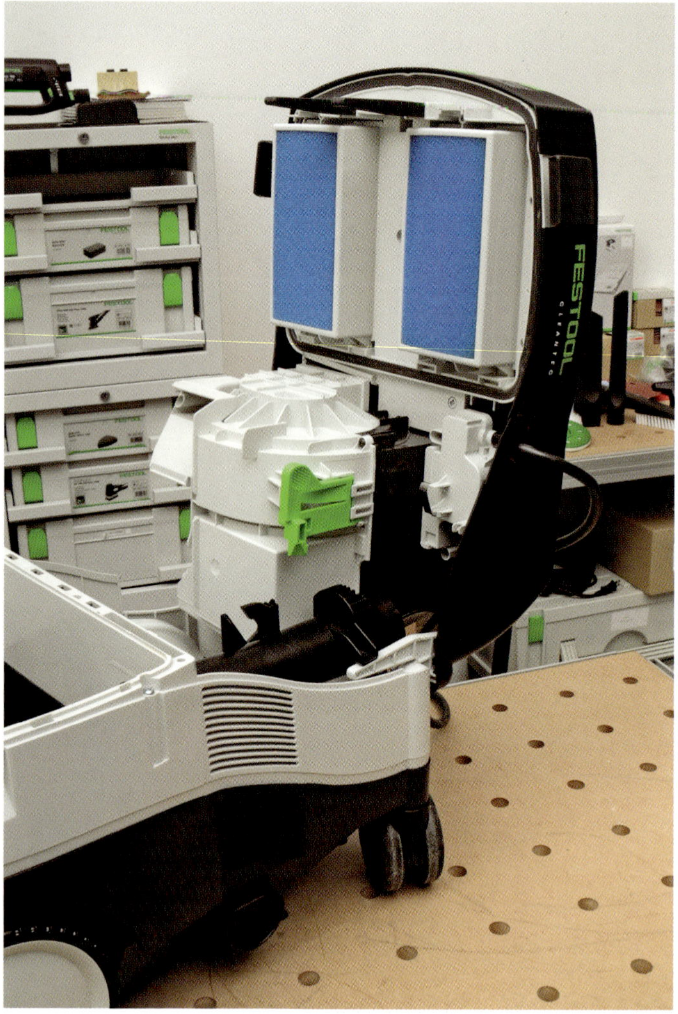

Pivot the top to the vertical position.

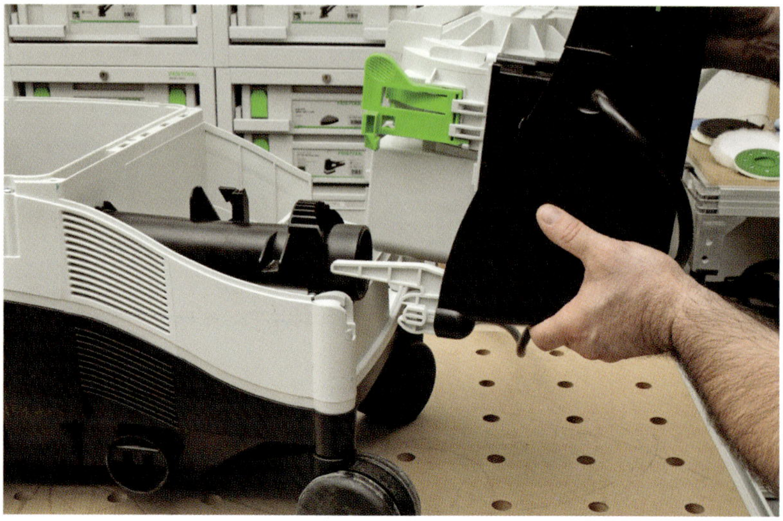

Pull the top straight out.

There is a handle at the front end...

...which makes emptying quite easy.

Reinsert the top. The long piece on the left goes above and the smaller rounder tab gets inserted into the slot.

Wet Vacuuming with the CT 22 E & CT 33 E

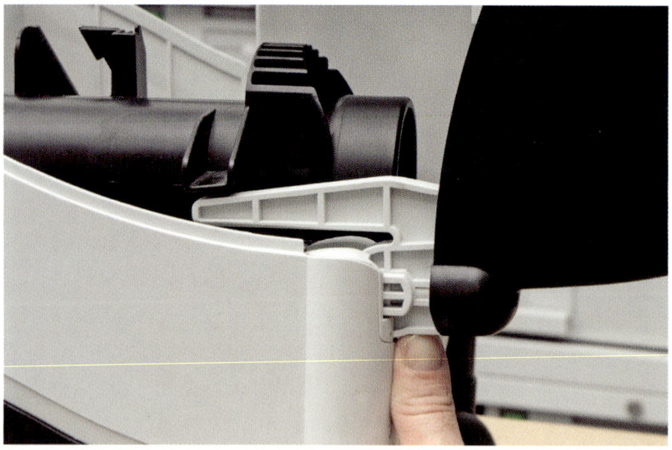

Progress

In place.

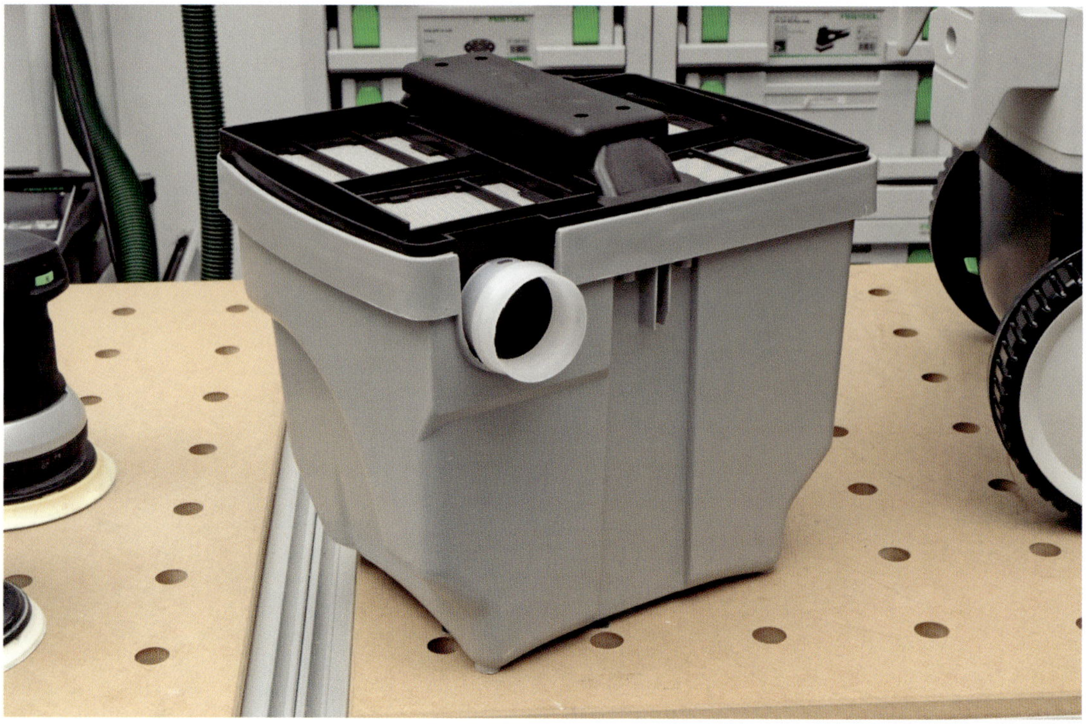

The Dirt Trap accessory provides an alternative for wet vac applications.

Wet Vacuuming with the CT 22 E & CT 33 E

Insert the pail into the well.

The dirt screen then fits on top. Align the input in the hole...

...and lower into place. The pail can simply be removed for emptying.

Hoses

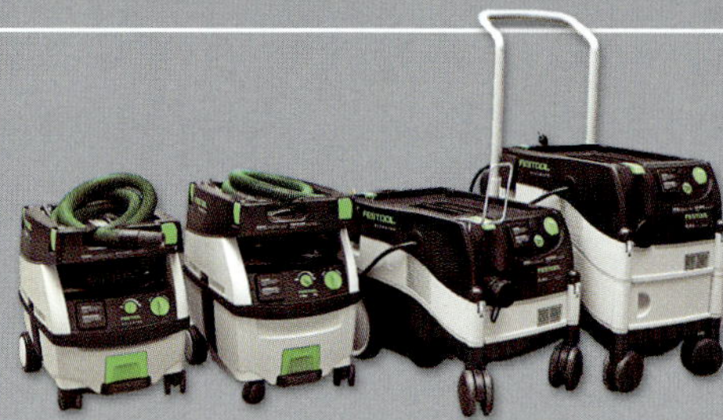

The standard hose for all dust extractors is antistatic, 27mm (1-1/16 inch) diameter, 3.5 meter (11.5 feet) long. This coordinates with the standard tool power cord, which is 13 feet long. It is also available in lengths up to 16.5 feet, which would require an extension cord.

A 36mm (1-7/16 inch) antistatic hose is available for the CT 22 E and CT 33 E. This will prevent clogging when using tools that produce larger chips and sawdust quickly, like a router or planer. Available in lengths of 11.5 feet, 16.5, and 23 feet.

Hoses

The dust shrouds of the tools are designed to fit with both the 27mm and the 36mm hoses. On the 27mm, the hose fits in the shroud.

On the 36mm, the shroud fits in the hose.

Hoses

The largest antistatic hose is 50mm (2-inches). It is offered in 8.25 and 13 feet. Its main purposes are to connect the dust extractor to shop tools and for shop clean-up.

All the hoses just discussed are available as non-antistatic, "gray" hose. This is good for the exhaust feature on the CT 22 E and CT 33 E, and for connection to a dust extractor by other manufacturers that do not offer the antistatic feature.

Connectors for the CT 22 E and CT 33 E

The Y-piece, allows the dust extractor to be connected to two devices. It plugs into the extractor, and has a blanking plug to close one of the branches when not in use.

Sometimes it is necessary to extend the hose. The combination depends on the application. In this case a 36mm hose from the extractor, on the right, is connecting to a 27mm hose going to the tool, on the left.

Connectors for the CT 22 E and CT 33 E

First, pull off the collar.

Then, unscrew the 36mm rotating connector collar. It follows the ribs in the hose, so it unscrews in a clockwise direction.

The rotating connector comes in antistatic and non-antistatic varieties. It has three parts as shown here. Left to right: rotating hose connector, threaded hose sleeve, and hose connector clip.

Connectors for the CT 22 E and CT 33 E

Place the clip around the hose.

Screw the hose sleeve onto the hose, turning it counterclockwise.

Clip the rotating connector in place.

Connectors for the CT 22 E and CT 33 E

Complete.

The two hoses can now be joined with the connector sleeve or to a hose reducer coupling sleeve as needed.

Connectors, left to right: Antistatic connector sleeve; antistatic rotating connector sleeve for 27mm suction hose; antistatic rotating connector sleeve for 36mm suction hose; antistatic rotating connector sleeve for 50mm suction hose.

Other Accessories

A blanking plug can be used to seal the intake hole in the dust extractor. The pointed tail…

…fits into a slot beneath the hole.

This is a convenient accessory for keeping the dust from escaping during transport.

The spark trap can be added to the system and is vital when working with metal. The baffled chamber cools down the metal particulate before it hits the bag. It helps to reduce the chance of fire from stray sparks while maintaining air flow.

The hose holder is inserted into the accessory socket at the upper front corner.

The hose holder gets the hose up and level with the work area.

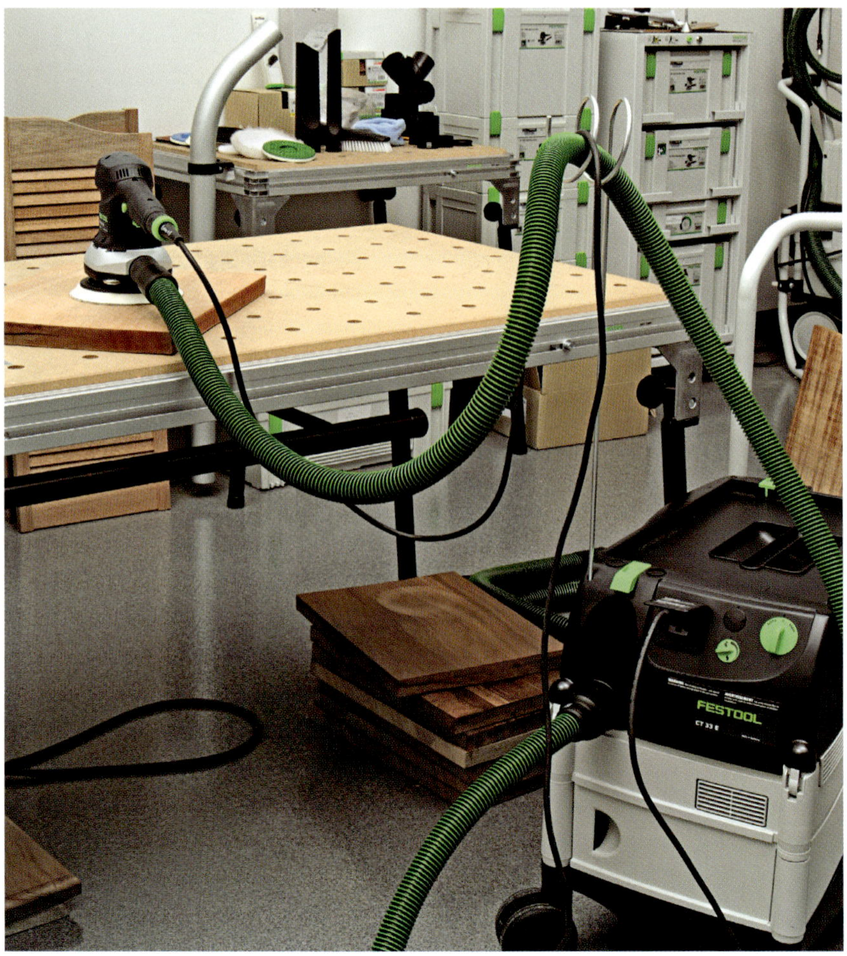

Other Accessories

Alternatively, the hose hook fits on the front on the CT22 or CT33, with the rod inserted in the socket at the front corner, and a screw holding it in place. It provides convenient hose storage.

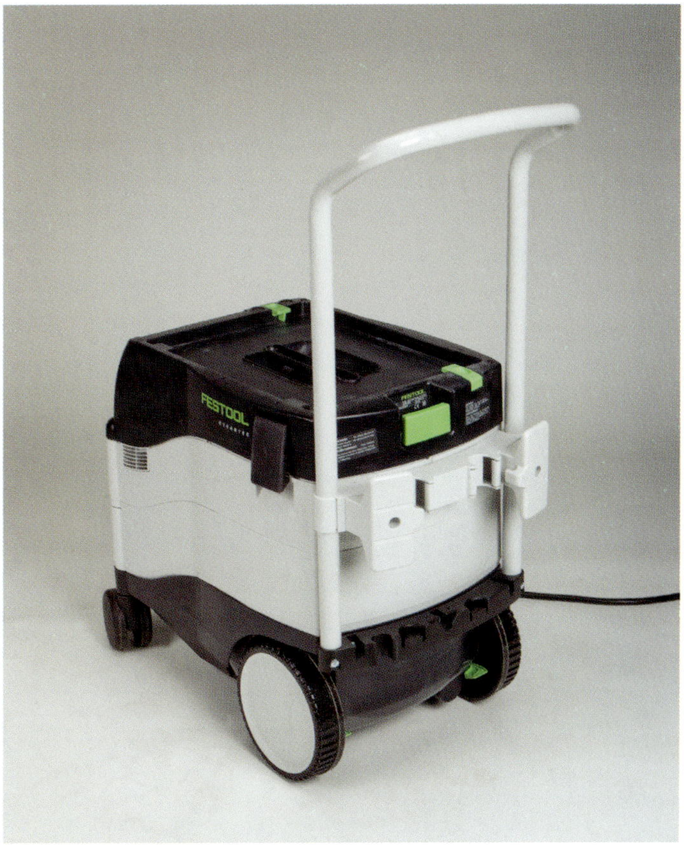

The optional handle assembly makes moving the CT 22 E or CT 33 E even easier. With it the unit can be pushed while standing as opposed to leaning over or carrying. When transporting several Systainers on the Sys-Dock feature, the handle converts the CT into a useful "dolly."

Other Accessories

The installation of the optional handle begins with the removal of the cord holders from the back of the machine.

Install the bottom handle assembly...

...and the upper handle assembly.

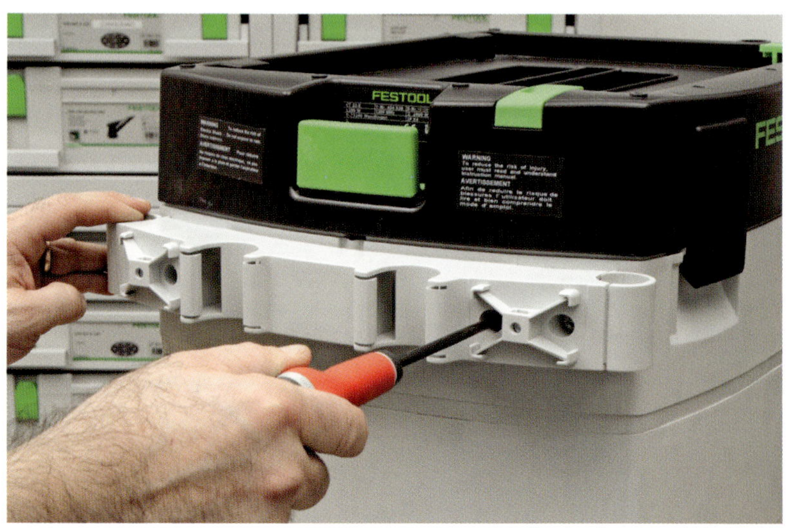

Insert a nut into the slot.

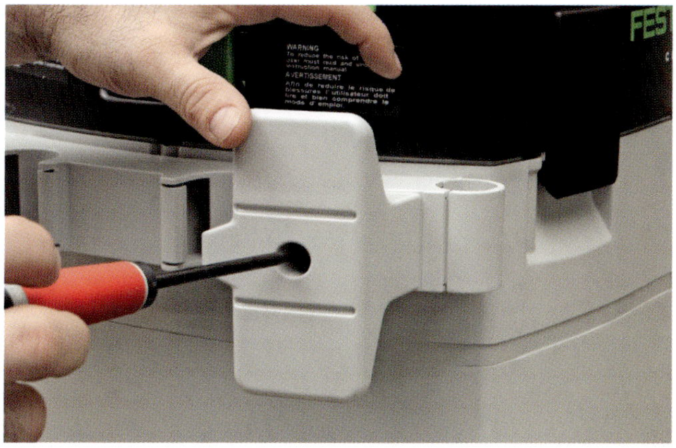

Reinstall the cord holders.

...and secure.

Insert the handle...

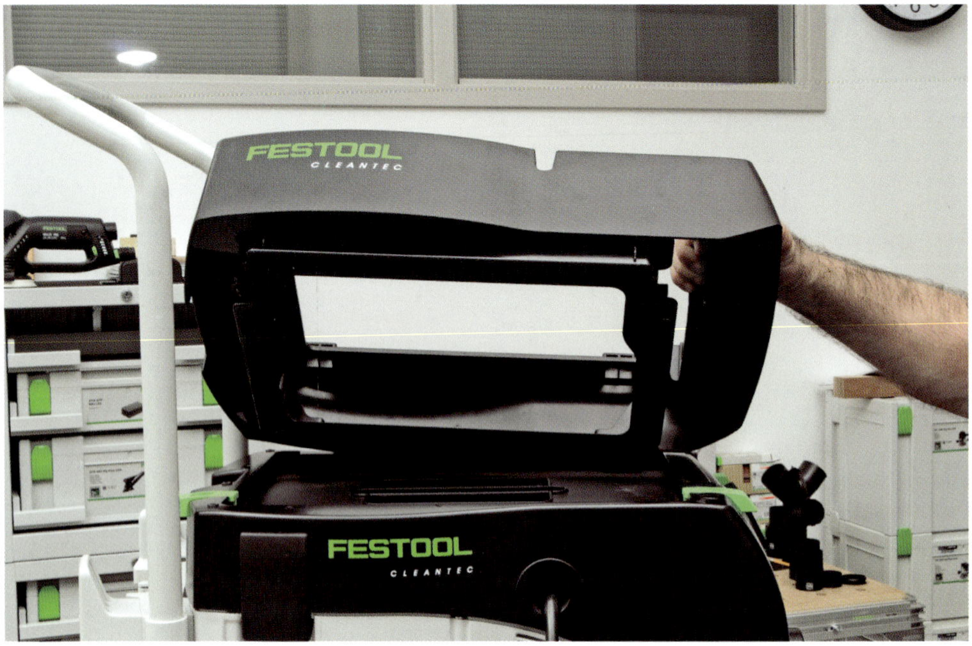

The CT hose garage fits on the top and provides convenient storage space for the hose and cord. Set it in place, with tabs in the slots along the left side…

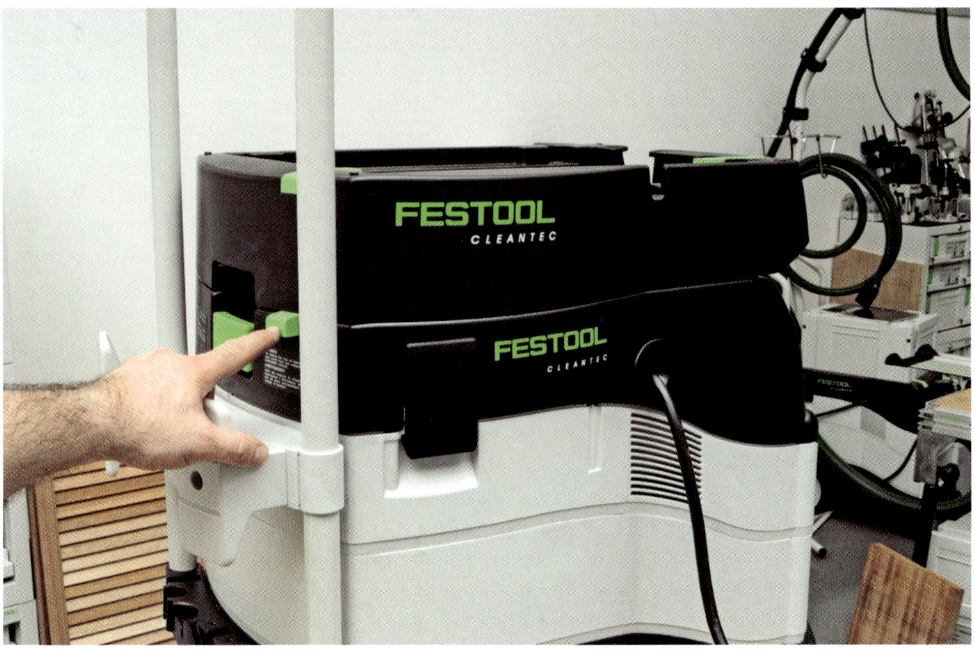

…and lock it.

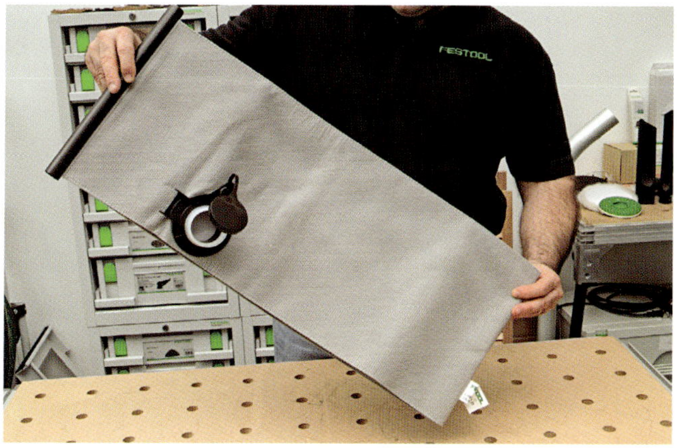

The Longlife filter bag provides a reusable alternative to paper.

It is lined with fabric for fine filtration.

It closes with this slider.

Place the bag in the well and connect to the dust extraction tube.

The Boom Arm for the CT 22 E and CT 33 E

The Boom Arm accessory keeps the dust extractor hose accessible to the work area, while lifting it up and out of the way of the work at hand. The pivoting arm moves with the tool so it doesn't hinder the work process. It works with the CT 22 E and the CT 33 E and requires the handle assembly. The boom arm comes with a 50mm antistatic extension hose and 3-wire 12 gauge extension power cord. It will accommodate a 27mm or 36mm tool hose, as well as the Festool pneumatic 3-in-1 hose for air tools.

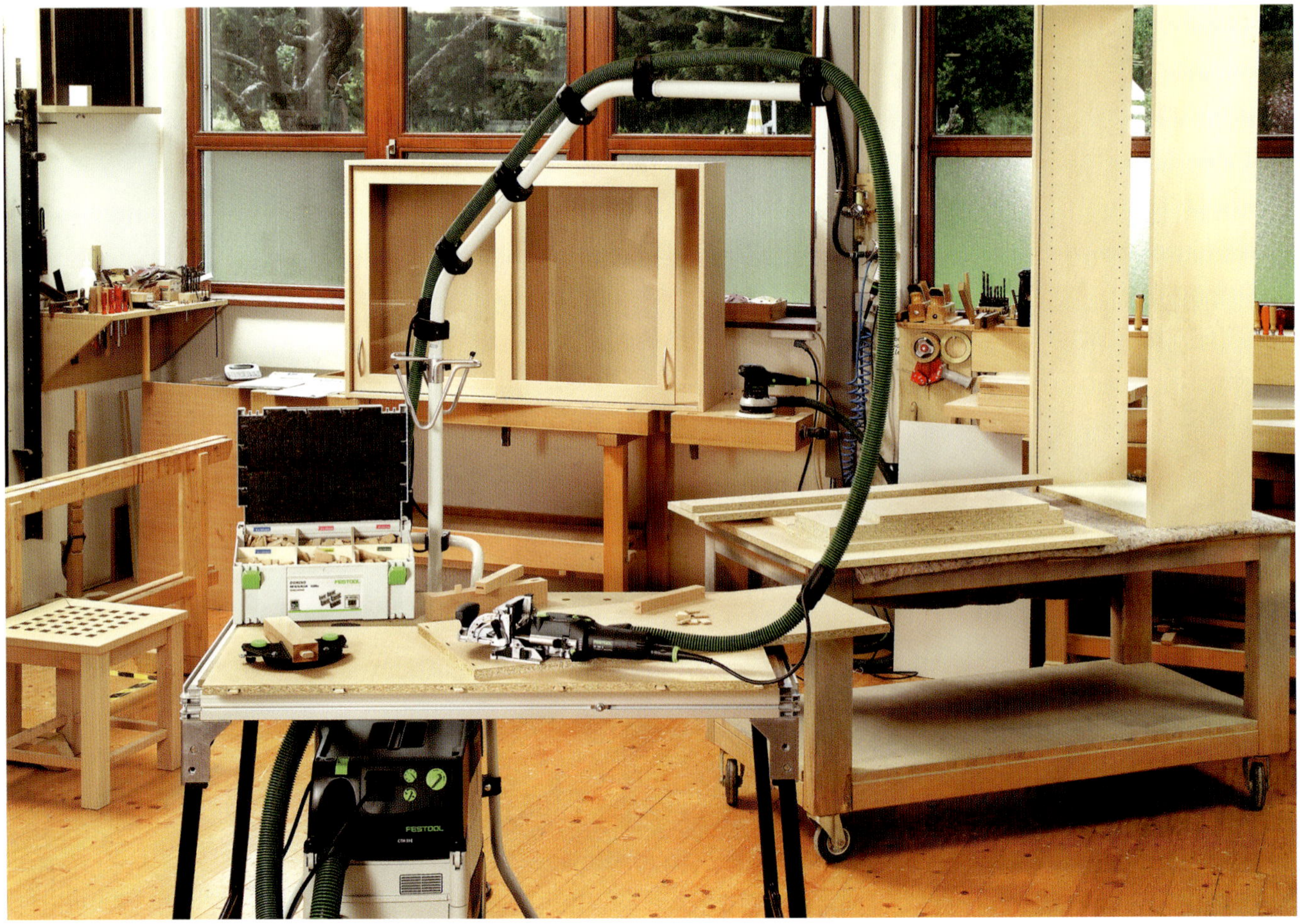

The Boom Arm for the CT 22 E and CT 33 E

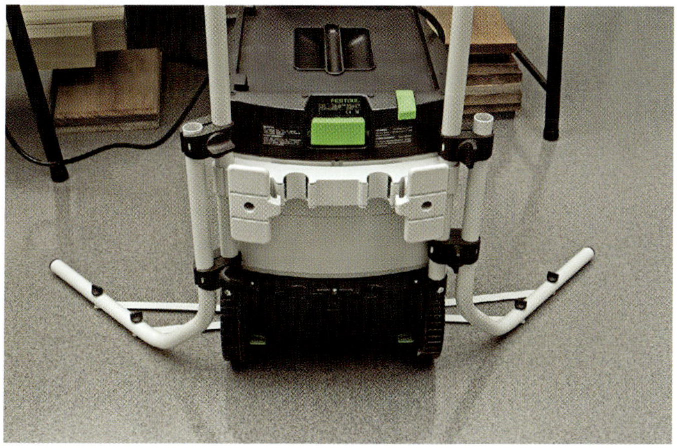

The boom arm requires the handle assembly to be in place. Attaching the boom begins with the outriggers, two added feet that give the unit stability. They are attached to the handles with two tool clamps on each side.

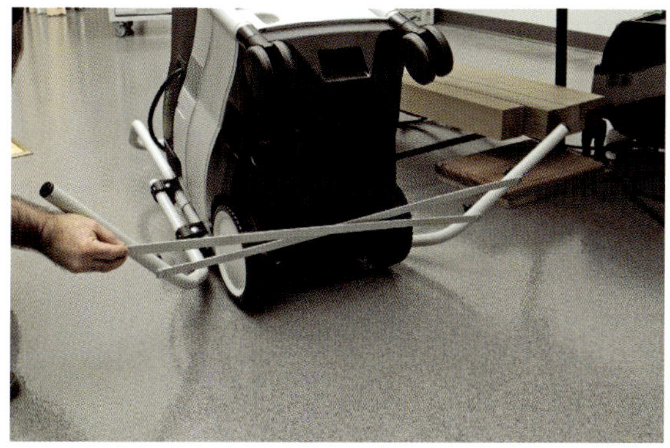

Cross braces give the outriggers added rigidity. Note: This configuration is good if you tend to run the dust extractor from the front of the machine. If you work more from the back the outriggers can be reversed.

The vertical fits on the lower handle bracket, where a tab keeps it from turning. It then fits in the upper handle support, where a screw clamp underneath, keeps it from riding up.

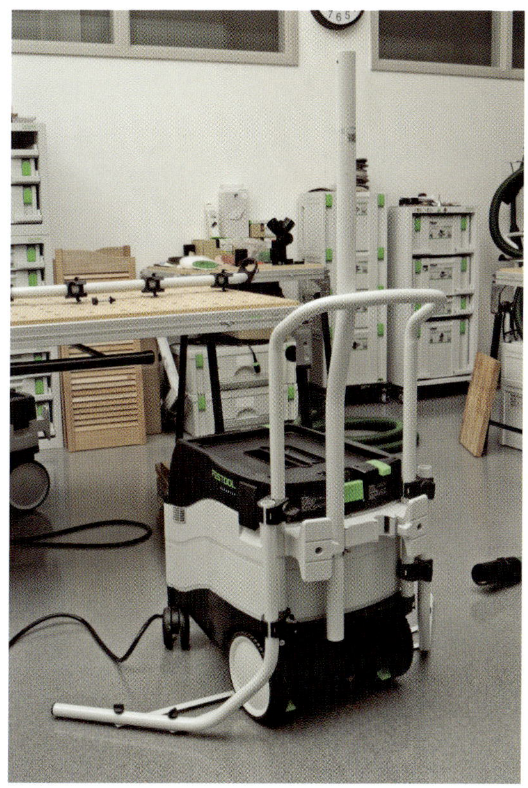

The vertical piece in place

The Boom Arm for the CT 22 E and CT 33 E

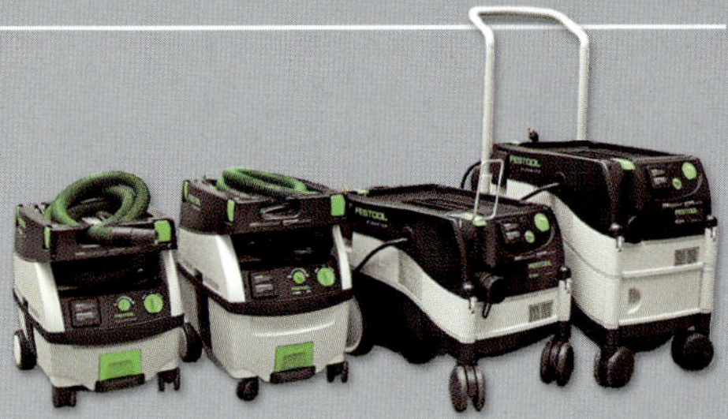

The vertical support is held to the handle with a clamp. The underside holds three recessed nuts.

Place the bottom piece around the vertical support and under the handle.

The Boom Arm for the CT 22 E and CT 33 E

Place the top piece in position.

Hold it in place with three bolts.

The Boom Arm for the CT 22 E and CT 33 E

The vertical support is secured to the bracket with a wire clamp, which is inserted into two holes at the back of the bracket.

In place.

The Boom Arm for the CT 22 E and CT 33 E

The arm arrives in two segments.

Snap them together.

The Boom Arm for the CT 22 E and CT 33 E

Insert the arm into the vertical support.

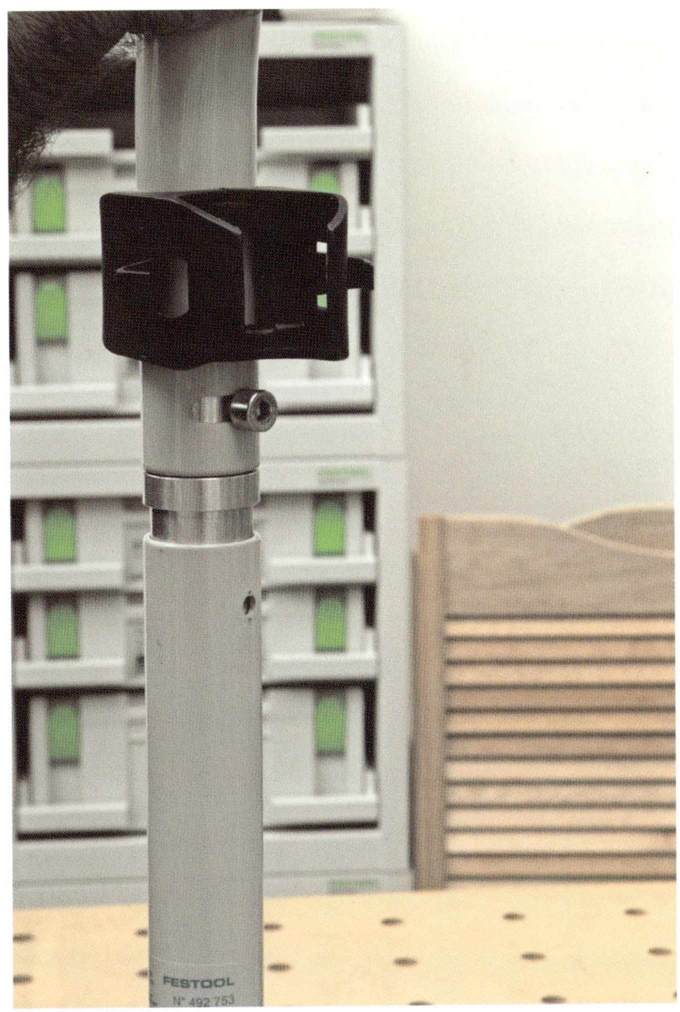

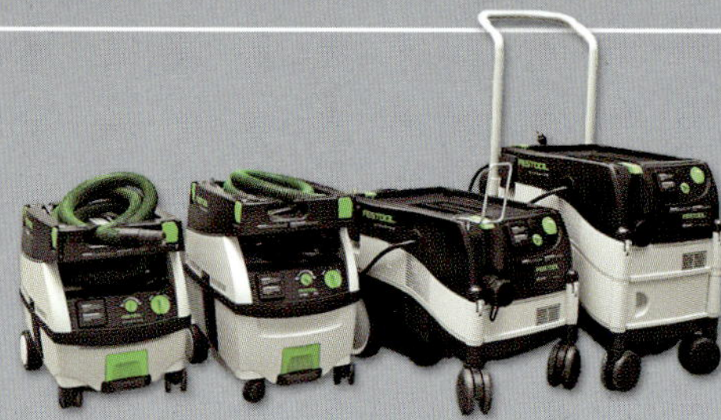

Secure with the bolt provided.

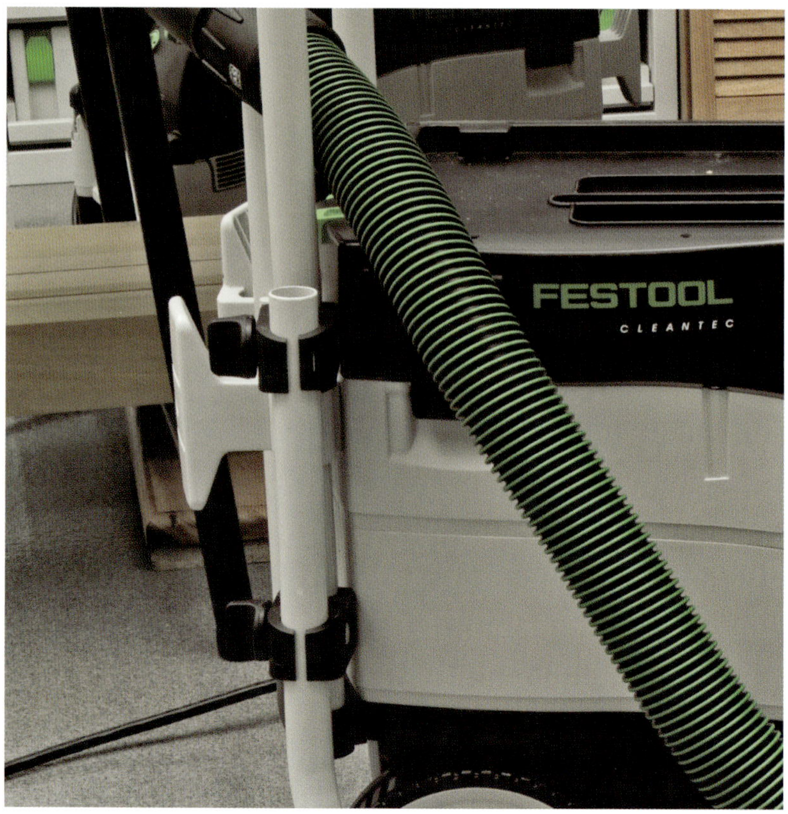

Ready for installation of the hose and power cord.

The Boom Arm for the CT 22 E and CT 33 E

The arm comes with a 50mm antistatic extension hose and 3-wire 12 gauge extension power cord, that run from the extractor to the boom arm.

The Boom Arm for the CT 22 E and CT 33 E

The arm will accommodate a 27mm or 36mm tool hose, as well as the Festool pneumatic 3-in-1 hose for air tools. Insert it in the hose brackets, starting where the arm joins the hose coming off the dust extractor.

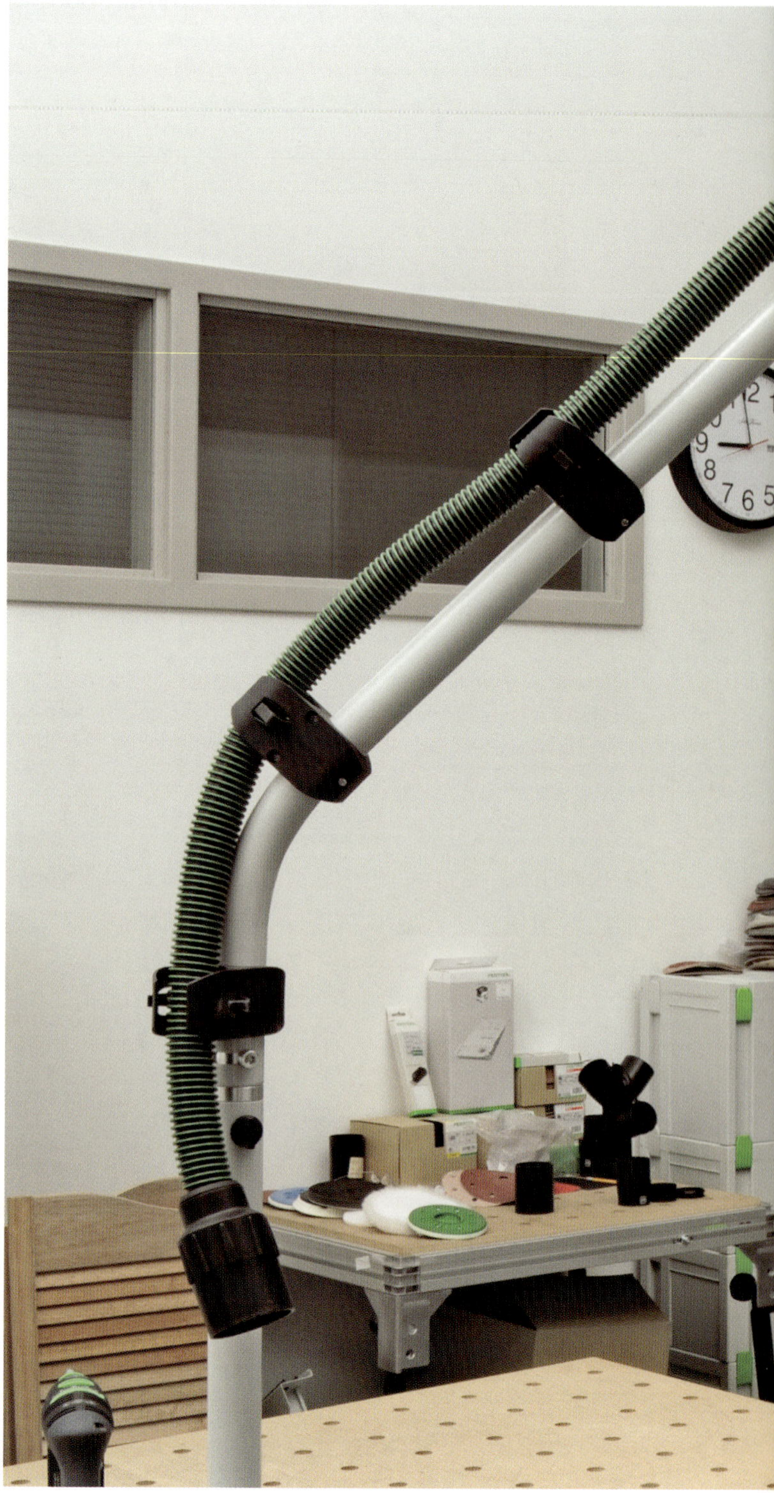

Continue to the end of the arm

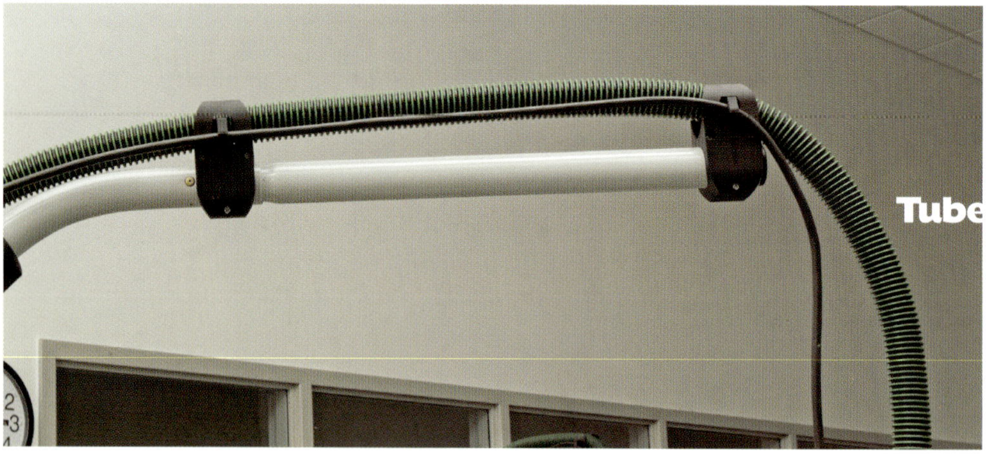

The tool's power cord is clamped to the outside of the hose brackets.

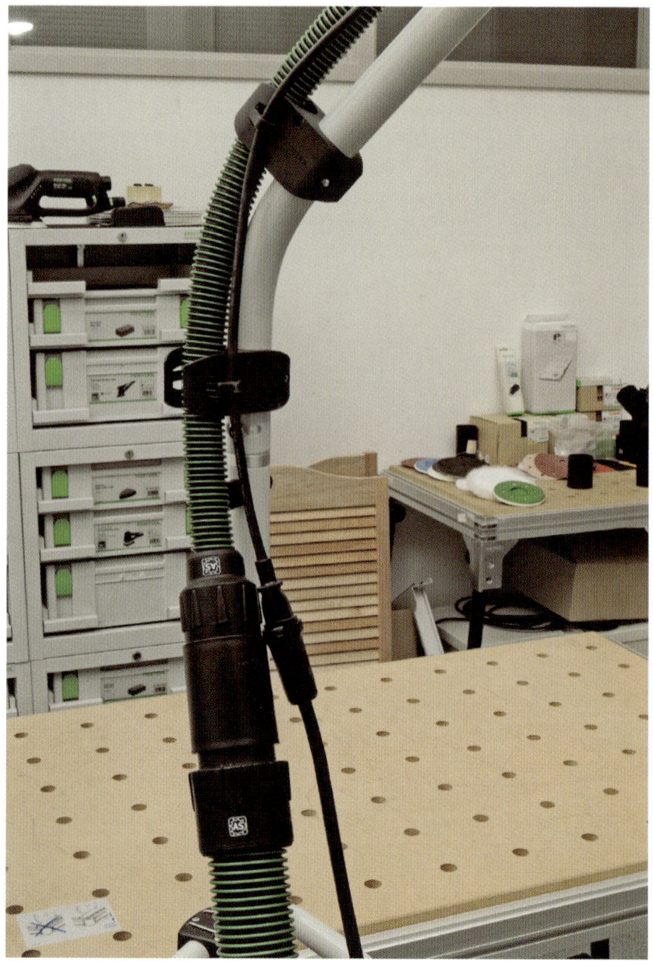

Connect the power cords and hoses.

The Boom Arm for the CT 22 E and CT 33 E

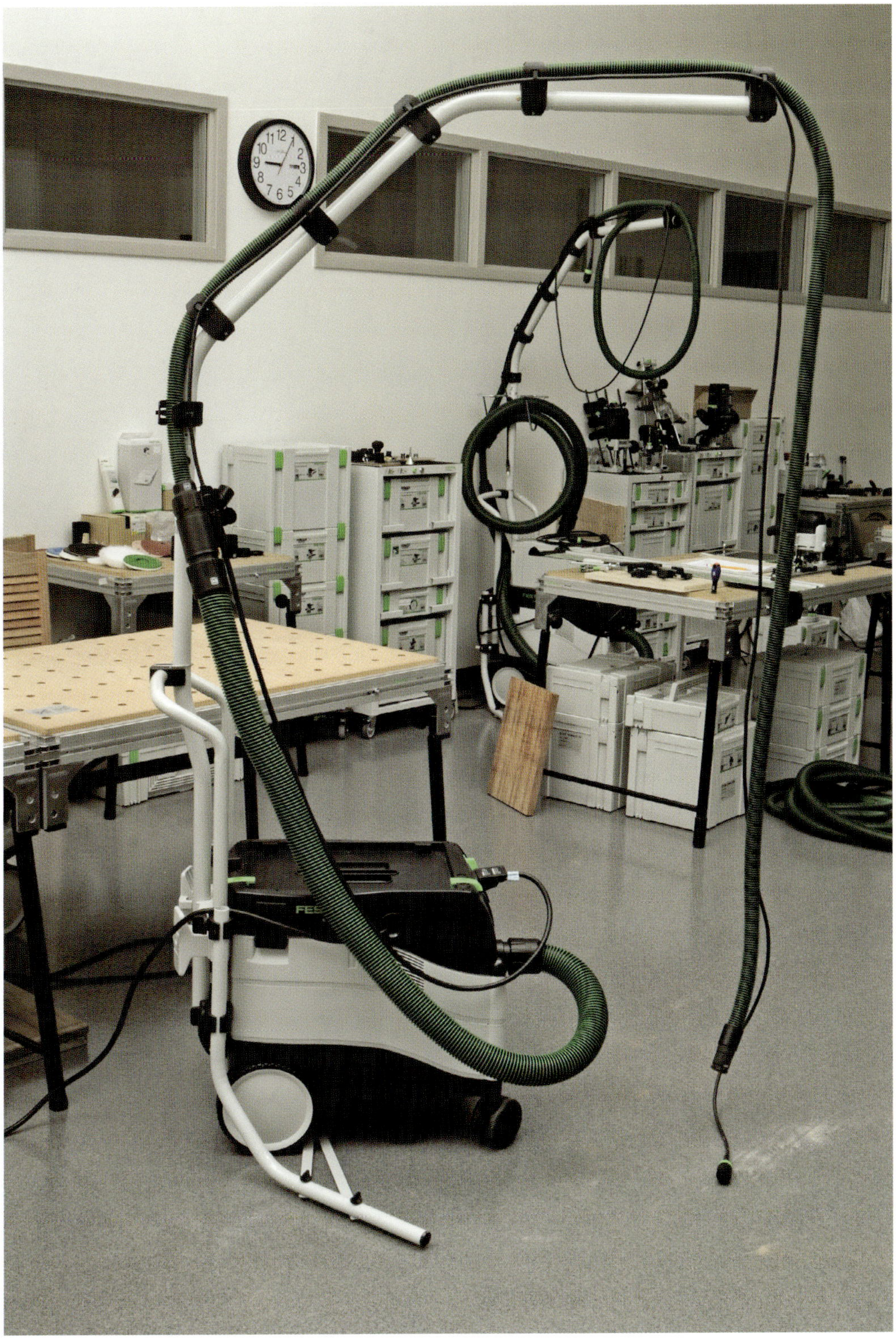

Complete

Tubes, Nozzles, & Brushes

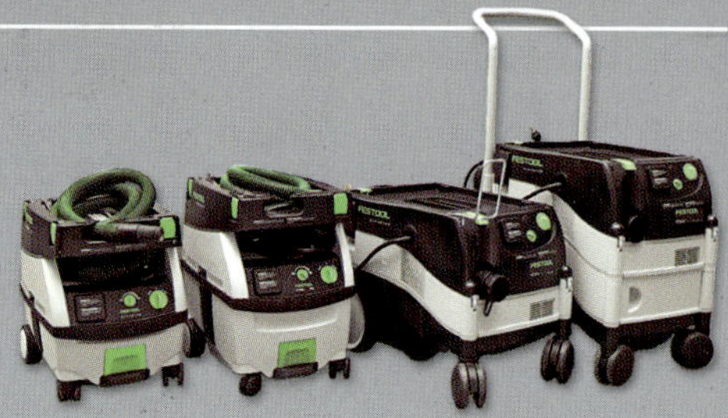

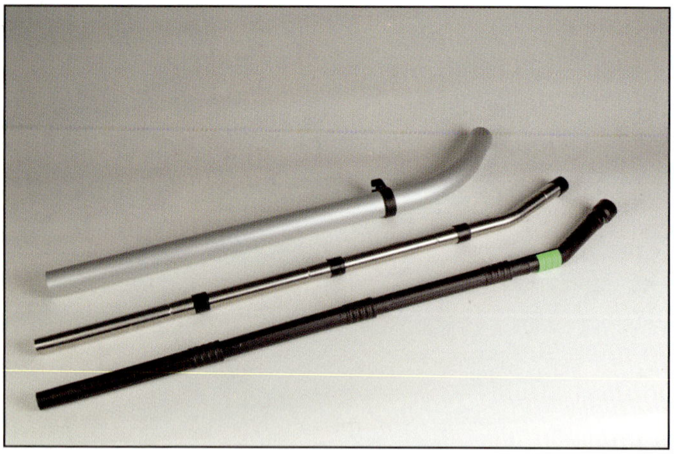

Top to bottom: one-piece anodized aluminum tub; stainless steel curved tube and extensions; curved polyproplyene (PP) curved tube with regulating slide to adjust suction power.

Nozzles & brushes (left to right): Large crevice nozzle; crevice nozzle; universal brush nozzle; upholstery brush; bevel ended nozzle; suction brush.

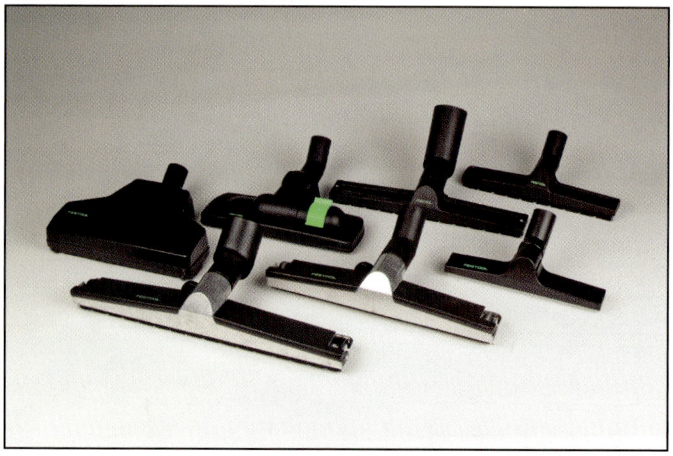

Floor nozzles (front row, left to right): Large industrial floor nozzle; industrial floor nozzle; interchangeable floor nozzle; (back row, left to right): turbo suction brush; multi-purpose floor nozzle; workshop floor nozzle; standard floor nozzle.